D0439178

PATTERNS OF PLAUSIBLE INFERENCE

MATHEMATICS
AND PLAUSIBLE REASONING

VOL. I. INDUCTION AND ANALOGY IN MATHEMATICS

VOL. II. PATTERNS OF PLAUSIBLE INFERENCE

Also by G. Polya:

HOW TO SOLVE IT

PATTERNS OF PLAUSIBLE INFERENCE

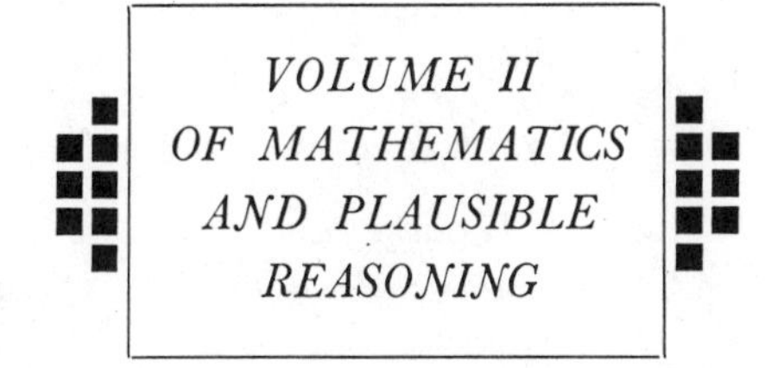

By G. POLYA

PRINCETON UNIVERSITY PRESS
PRINCETON, NEW JERSEY
1968

Published by Princeton University Press, 41 William Street,
Princeton, New Jersey 08540

Library of Congress Card No. 68-56327
ISBN 0-691-08006-2 (cloth)
ISBN 0-691-02510-X (paper)
First Princeton Paperback printing, 1990

Princeton University Press books are printed on acid-free paper, and meet the guidelines for permanence and durability of the Committee on Production Guidelines for Book Longevity of the Council on Library Resources

15 14 13 12 11 (cloth edn.)
10 9 8 7 6 5 4 3 2 (paper edn.)

Printed in the United States of America
by Princeton University Press,
Princeton, New Jersey

PREFACE

Inductive reasoning is one of the many battlefields for conflicting philosophical opinions, and one that is still relatively lively today. The reader who went through Vol. I of this work has had a good opportunity to notice two things. First, inductive and analogical reasoning play a major rôle in mathematical discovery. Second, both inductive and analogical reasoning are particular cases of plausible reasoning. It seems to me more philosophical to consider the general idea of plausible reasoning instead of its isolated particular cases. The present Vol. II attempts to formulate certain patterns of plausible reasoning, to investigate their relation to the Calculus of Probability, and to examine in what sense they can be regarded as "rules" of plausible reasoning. Their relation to mathematical invention and instruction will also be briefly discussed.

The text of the present Vol. II does not often refer explicitly to Vol. I, and the reader can understand the main connections without looking up these references. Among the problems appended to the various chapters there are some that the reader cannot solve without referring to Vol. I, but on the whole one can read Vol. II in first approximation without having read Vol. I. Yet, of course, it is more natural to read Vol. II after Vol. I, the examples of which provide the investigation that lies ahead of us with experimental data and a richer background.

Such data and background are particularly desirable in view of the method that will be followed. I wish to investigate plausible reasoning in the manner of the naturalist: I collect observations, state conclusions, and emphasize the points in which my observations seem to support my conclusions. Yet I respect the judgment of the reader and I do not want to force or trick him into adopting my conclusions.

Of course, the views presented here have no pretension to be final. In fact, there are a few places where I feel clearly the need of some improvement, minor or major. Yet I believe that the main direction is right, and that the discussions, and especially the examples, of this work may elucidate the "double nature" and the "complementary aspects" of plausible and especially inductive reasoning, which appears sometimes as "objective" and sometimes as "subjective."

George Polya

Stanford University
May 1953

PREFACE TO THE SECOND EDITION

The text of the first edition is reprinted with a few minor changes and an appendix is added.

The appendix contains an article, twelve comments, and thirty-three problems with solutions. The article "Heuristic reasoning in the theory of numbers" appeared previously in the *American Mathematical Monthly;* I wish to express my thanks to the Mathematical Association of America for the permission to reprint it. The comments and problems supplement various chapters in both volumes of this work.

GEORGE POLYA

Stanford University
May 1968

HINTS TO THE READER

THE section 2 of chapter VII is quoted as sect. 2 in chapter VII, but as sect. 7.2 in any other chapter. The subsection (3) of section 5 of chapter XIV is quoted as sect. 5 (3) in chapter XIV, but as sect. 14.5 (3) in any other chapter. We refer to example 26 of chapter XIV as ex. 26 in the same chapter, but as ex. 14.26 in any other chapter.

Some knowledge of elementary algebra and geometry may be enough to read substantial parts of the text. Thorough knowledge of elementary algebra and geometry and some knowledge of analytic geometry and calculus, including limits and infinite series, is sufficient for almost the whole text and the majority of the examples and comments. Yet more advanced knowledge is supposed in a few incidental remarks of the text, in some proposed problems, and in several comments. Usually some warning is given when more advanced knowledge is assumed.

The advanced reader who skips parts that appear to him too elementary may miss more than the less advanced reader who skips parts that appear to him too complex.

Some details of (not very difficult) demonstrations are often omitted without warning. Duly prepared for this eventuality, a reader with good critical habits need not spoil them.

Some of the problems proposed for solution are very easy, but a few are pretty hard. Hints that may facilitate the solution are enclosed in square brackets [ ]. The surrounding problems may provide hints. Especial attention should be paid to the introductory lines prefixed to the examples in some chapters, or prefixed to the First Part, or Second Part, of such examples.

The solutions are sometimes very short: they suppose that the reader has earnestly tried to solve the problem by his own means before looking at the printed solution.

A reader who spent serious effort on a problem may profit by it even if he does not succeed in solving it. For example, he may look at the solution, try to isolate what appears to him the key idea, put the book aside, and then try to work out the solution.

At some places, this book is lavish of figures or in giving small intermediate steps of a derivation. The aim is to render visible the *evolution* of a figure or a formula; see, for instance, fig. 16.1–16.5. Yet no book can have enough figures or formulas. A reader may want to read a passage "in first approximation" or more thoroughly. If he wants to read more thoroughly, he should have paper and pencil at hand: he should be prepared to write or draw any formula or figure given in, or only indicated by, the text. Doing so, he has a better chance to see the evolution of the figure or formula, to understand how the various details contribute to the final product, and to remember the whole thing.

CONTENTS

Volume II

Patterns of Plausible Inference

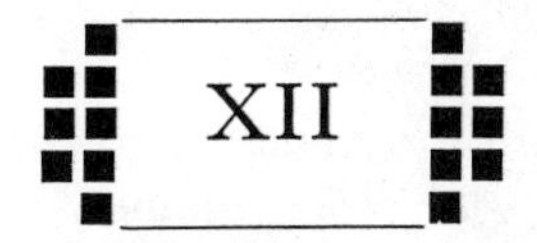

SOME CONSPICUOUS PATTERNS

I do not wish, at this stage, to examine the logical justification of this form of argumentation; for the present, I am considering it as a practice, *which we can observe in the habits of men and animals.*—BERTRAND RUSSELL[1]

1. Verification of a consequence. In the first volume of this work on *Induction and Analogy in Mathematics* we found some opportunity to familiarize ourselves with the practice of plausible reasoning. In the present second volume we undertake to describe this practice in general terms. The examples of the first part have already indicated certain forms or *patterns* of plausible reasoning. In the present chapter we undertake to formulate some such patterns explicitly.[2]

We begin with a pattern of plausible inference which is of so general use that we could extract it from almost any example. Yet let us take an example which we have not yet discussed before.

The following conjecture is due to Euler:[3] *Any integer of the form* $8n + 3$ *is the sum of a square and of the double of a prime.* Euler could not prove this conjecture, and the difficulty of a proof appears perhaps even greater today than in Euler's time. Yet Euler verified his statement for all integers of the form $8n + 3$ under 200; for $n = 1, 2, \ldots 10$ see Table I.

Table I

$$11 = 1 + 2 \times 5$$
$$19 = 9 + 2 \times 5$$
$$27 = 1 + 2 \times 13$$
$$35 = 1 + 2 \times 17 = 9 + 2 \times 13 = 25 + 2 \times 5$$
$$43 = 9 + 2 \times 17$$
$$51 = 25 + 2 \times 13$$
$$59 = 1 + 2 \times 29 = 25 + 2 \times 17 = 49 + 2 \times 5$$
$$67 = 9 + 2 \times 29$$
$$75 = 1 + 2 \times 37 = 49 + 2 \times 13$$
$$83 = 1 + 2 \times 41 = 9 + 2 \times 37 = 25 + 2 \times 29 = 49 + 2 \times 17$$

[1] *Philosophy*, W. W. Norton & Co., 1927, p. 80.

[2] Parts of this chapter were used in my address "On plausible reasoning" printed in the *Proceedings of the International Congress of Mathematicians* 1950, vol. 1, p. 739–747.

[3] *Opera Omnia*, ser. 1, vol. 4, p. 120–124. In this context, Euler regards 1 as a prime; this is needed to account for the case $3 = 1 + 2 \times 1$.

Such empirical work can be easily carried further; no exception has been found in numbers under 1000.[4] Does this prove Euler's conjecture? By no means; even verification up to 1,000,000 would prove nothing. Yet each verification renders the conjecture somewhat more credible, and we can see herein a general pattern.

Let A denote some clearly formulated conjecture which is, at present, neither proved, nor disproved. (For instance, A may be Euler's conjecture that, for $n = 1, 2, 3, \ldots$,

$$8n + 3 = x^2 + 2p$$

where x is an integer and p a prime.) Let B denote some consequence of A; also B should be clearly stated and neither proved, nor disproved. (For instance, B may be the first particular case of Euler's conjecture not listed in Table I which asserts that $91 = x^2 + 2p$.) For the moment we do not know whether A or B is true. We do know, however, that

A implies B.

Now, we undertake to check B. (A few trials suffice to find out whether the assertion about 91 is true or not.) If it turned out that B is false, we could conclude that A also is false. This is completely clear. We have here a classical elementary pattern of reasoning, the "modus tollens" of the so-called hypothetical syllogism:

$$\frac{\begin{array}{c} A \text{ implies } B \\ B \text{ false} \end{array}}{A \text{ false}}$$

The horizontal line separating the two premises from the conclusion stands as usual for the word "therefore." We have here *demonstrative inference* of a well-known type.

What happens if B turns out to be true? (Actually, $91 = 9 + 2 \times 41 = 81 + 2 \times 5$.) There is no demonstrative conclusion: the verification of its consequence B does not prove the conjecture A. Yet such verification renders A more credible. (Euler's conjecture, verified in one more case, becomes somewhat more credible.) We have here a pattern of *plausible inference*:

$$\frac{\begin{array}{c} A \text{ implies } B \\ B \text{ true} \end{array}}{A \text{ more credible}}$$

The horizontal line again stands for "therefore." We shall call this pattern the *fundamental inductive pattern* or, somewhat shorter, the "inductive pattern."

[4] Communication of Professor D. H. Lehmer.

This inductive pattern says nothing surprising. On the contrary, it expresses a belief which no reasonable person seems to doubt: *The verification of a consequence renders a conjecture more credible.* With a little attention, we can observe countless reasonings in everyday life, in the law courts, in science, etc., which appear to conform to our pattern.

2. Successive verification of several consequences. In the present section, I use the phrase "discussion of a theorem" in the specific meaning "discussion, or survey, of some particular cases and some more immediate consequences of the theorem." I think that the discussion of the theorems presented is useful both in advanced and in elementary classes. Let us consider a very elementary example. Let us assume that you teach a class in solid geometry and that you have to derive the formula for the area of the lateral surface of the frustum of a cone. Of course, the cone is a right circular cone, and you are given the radius of the base R, the radius of the top r, and the altitude h. You go through the usual derivation and you arrive at the result:

A. The area of the lateral surface of the frustum is

$$\pi(R+r)\sqrt{(R-r)^2+h^2}.$$

We call this theorem A for future reference.

Now comes the discussion of the theorem A. You ask the class: *Can you check the result?* If there is no response, you give more explicit hints: Can you check the result by *applying* it? Can you check it by applying it to some *particular case you already know*? Eventually, with more or less collaboration from the part of your class, you get down to various known cases. If $R = r$, you obtain a first noteworthy particular case:

B_1. *The area of the lateral surface of a cylinder is* $2\pi rh$.

Of course, h stands for the altitude of the cylinder and r for the radius of its base. We call B_1 this consequence of A for future reference. The consequence B_1 has been treated already in your class and so it serves as a confirmation of A.

You obtain another particular case of A in setting $r = 0$ which yields:

B_2. *The area of the lateral surface of a cone is* $\pi R\sqrt{R^2+h^2}$.

Here h denotes the altitude of the cone and R the radius of its base. Also this consequence B_2 of A was known before and serves as a further confirmation of A.

There is a less obvious but interesting particular case corresponding to $h = 0$:

B_3. *The area of the annulus between two concentric circles with radii R and r is* $\pi R^2 - \pi r^2$.

This consequence B_3 of A is clear from plane geometry and yields still another confirmation of A.

The foregoing three particular cases, all known from previous study, support A from three different sides; the three figures (cylinder, cone, and annulus, corresponding to $r = R$, $r = 0$, and $h = 0$, respectively) look quite different. You may mention also the very particular case $r = h = 0$.

B_4. *The area of a circle with radius R is* πR^2.

I have sometimes observed that a boy in the last row who seemed to sleep soundly toward the end of my careful derivation opened his eyes and showed some interest in the progress of the discussion. The derivation of the formula, apparently plain and easy, seemed abstruse and difficult to him. He was not convinced by the derivation. He is more convinced by the discussion: a formula that checks in so many and so different cases has a good chance to be correct, he thinks. And in thinking so he conforms to a pattern of plausible reasoning which is closely related to, but more sophisticated than, the fundamental inductive pattern:

$$\begin{array}{c} A \text{ implies } B_{n+1} \\ B_{n+1} \text{ is very different from the formerly} \\ \text{verified consequences } B_1, B_2, \ldots B_n \text{ of } A \\ B_{n+1} \text{ true} \\ \hline A \text{ much more credible} \end{array}$$

This pattern adds a qualification to the fundamental inductive pattern. Certainly the verification of any consequence strengthens our belief in a conjecture. Yet the verification of certain consequences strengthens our belief more and that of others strengthens it less. The pattern just given brings to our attention a circumstance which has a great influence on the strength of inductive evidence: the variety of the consequences tested. The verification of a new consequence counts more if the new consequence differs more from the formerly verified consequences.

Now let us look at the reverse of the medal. Take the example of the foregoing section 1. The successive cases in Table I in which Euler's conjecture is verified look very similar to each other—unless we notice some hidden clue, and it seems very difficult to notice such a clue. Therefore, sooner or later, we get tired of this monotonous sequence of verifications. Having verified a certain number of cases, we hesitate. Is it worth while to tackle the next case? The next case, if the result is negative, could explode the conjecture—but the next case is so similar in all known aspects to the cases already verified that we scarcely expect a negative result. The next case, if the result is positive, would increase our confidence in Euler's conjecture, but this increase in confidence would be so small that it is scarcely worth the trouble of testing that next case.

This consideration suggests the following pattern which is not essentially different from the pattern that we have just stated, but rather a complementary form of it:

$$\begin{array}{c} A \text{ implies } B_{n+1} \\ B_{n+1} \text{ is very similar to the formerly verified} \\ \text{consequences } B_1, B_2, \ldots B_n \text{ of } A \\ B_{n+1} \text{ true} \\ \hline A \text{ just a little more credible} \end{array}$$

The verification of a new consequence counts more or less according as the new consequence differs more or less from the formerly verified consequences.

3. Verification of an improbable consequence. In a little known short note[5] Euler considers, for positive values of the parameter n, the series

$$(1)\quad 1 - \frac{x^2}{n(n+1)} + \frac{x^4}{n(n+1)(n+2)(n+3)} - \frac{x^6}{n\ldots(n+5)} + \ldots$$

which converges for all values of x. He observes the sum of the series and its zeros for $n = 1, 2, 3, 4$.

$n = 1$: sum $\cos x$, zeros $\pm \pi/2, \pm 3\pi/2, \pm 5\pi/2, \ldots$

$n = 2$: sum $(\sin x)/x$, zeros $\pm \pi, \pm 2\pi, \pm 3\pi, \ldots$

$n = 3$: sum $2(1 - \cos x)/x^2$, zeros $\pm 2\pi, \pm 4\pi, \pm 6\pi, \ldots$

$n = 4$: sum $6(x - \sin x)/x^3$, no real zeros.

Euler observes a difference: in the first three cases all the zeros are real, in the last case none of the zeros is real. Euler notices a more subtle difference between the first two cases and the third case: for $n = 1$ and $n = 2$, the distance between two consecutive zeros is π (provided that we disregard the zeros next to the origin in the case $n = 2$) but for $n = 3$ the distance between consecutive zeros is 2π (with a similar proviso). This leads him to a striking observation: in the case $n = 3$ all the zeros are double zeros. "Yet we know from Analysis," says Euler, "that two roots of an equation always coincide in the transition from real to imaginary roots. Thus we may understand why all the zeros suddenly become imaginary when we take for n a value exceeding 3." On the basis of these observations he states a surprising conjecture: the function defined by the *series* (1) *has only real zeros, and an infinity of them, when* $0 < n \leq 3$, *but has no real zero at all when* $n > 3$. In this statement he regards n as a continuously varying parameter.

In Euler's time questions about the reality of the zeros of transcendental equations were absolutely new, and we must confess that even today we possess no systematic method to decide such questions. (For instance, we

[5] *Opera Omnia*, ser. 1, vol. 16, sect. 1, p. 241–265.

cannot prove or disprove Riemann's famous hypothesis.) Therefore, Euler's conjecture appears extremely bold. I think that the courage and clearness with which he states his conjecture are admirable.

Yet Euler's admirable performance is understandable to a certain extent. Other experts perform similar feats in dealing with other subjects, and each of us performs something similar in everyday life. In fact, Euler *guessed the whole from a few scattered details.* Quite similarly, an archaeologist may reconstitute with reasonable certainty a whole inscription from a few scattered letters on a worn-out stone. A paleontologist may describe reliably the whole animal after having examined a few of its petrified bones. When a person whom you know very well starts talking in a certain way, you may predict after a few words the whole story he is going to tell you. Quite similarly, Euler guessed the whole story, the whole mathematical situation, from a few clearly recognized points.

It is still remarkable that he guessed it from so few points, by considering just four cases, $n = 1, 2, 3, 4$. We should not forget, however, that circumstantial evidence may be very strong. A defendant is accused of having blown up the yacht of his girl friend's father, and the prosecution produces a receipt signed by the defendant acknowledging the purchase of such and such an amount of dynamite. Such evidence strengthens the prosecution's case immensely. Why? Because the purchase of dynamite by an ordinary citizen is a very unusual event in itself, but such a purchase is completely understandable if the purchaser intends to blow up something or somebody. Please observe that this court case is very similar to the case $n = 3$ of Euler's series. That all roots of an equation written at random turn out to be double roots is a very unusual event in itself. Yet it is completely understandable that in the transition from two real roots to two imaginary roots a double root appears. The case $n = 3$ is the strongest piece of circumstantial evidence produced by Euler and we can perceive herein a general pattern of plausible inference:

$$\frac{\begin{array}{c} A \text{ implies } B \\ B \text{ very improbable in itself} \\ B \text{ true} \end{array}}{A \text{ very much more credible}}$$

Also this pattern appears as a modification or a sophistication of the fundamental inductive pattern (sect. 1). Let us add, without specific illustration for the moment, the complementary pattern which explains the same idea from the reverse side:

$$\frac{\begin{array}{c} A \text{ implies } B \\ B \text{ quite probable in itself} \\ B \text{ true} \end{array}}{A \text{ just a little more credible}}$$

The verification of a consequence counts more or less according as the consequence is more or less improbable in itself. The verification of the most surprising consequences is the most convincing.

By the way, Euler was right: 150 years later, his conjecture has been completely proved.[6]

4. Inference from analogy. At this stage it may be instructive to look back at the examples of the first volume on *Induction and Analogy.* We have formulated a few patterns of plausible inference in the foregoing sections of this chapter: how do those examples appear in the light of these patterns?

Let us reconsider two related examples (from sect. 10.1 and 10.4 of Volume I, respectively). One of these examples is connected with the isoperimetric theorem and Descartes, the other with a physical analogue of the isoperimetric theorem and Lord Rayleigh. We reproduce two tables from ch. X (called there Table I and Table II, here Table II and Table III, respectively) putting them side by side. Table II (as it is numbered in the present chapter) lists the perimeters of ten figures, each of which has the same area 1, and Table III lists the principal frequencies of the same ten figures (considered as vibrating membranes).

Table II
Perimeters
of Figures of Equal Area

Circle	3.55
Square	4.00
Quadrant	4.03
Rectangle 3 : 2	4.08
Semicircle	4.10
Sextant	4.21
Rectangle 2 : 1	4.24
Equilateral triangle	4.56
Rectangle 3 : 1	4.64
Isosceles right triangle	4.84

Table III
Principal Frequencies
of Membranes of Equal Area

Membrane	Frequency
Circle	4.261
Square	4.443
Quadrant	4.551
Sextant	4.616
Rectangle 3 : 2	4.624
Equilateral triangle	4.774
Semicircle	4.803
Rectangle 2 : 1	4.967
Isosceles right triangle	4.967
Rectangle 3 : 1	5.736

The perimeters in one table, the principal frequencies in the other, are increasingly ordered. Both tables start with the circle which has the shortest perimeter among the ten figures listed and also the lowest principal frequency, and this suggests two theorems:

Of all plane figures with a given area the circle has the shortest perimeter.

Of all membranes with a given area the circle has the lowest principal frequency.

[6] See the author's paper: Sopra una equazione transcendente trattata da Eulero, *Bolletino dell' Unione Matematica Italiana,* vol. 5, 1926, p. 64–68.

The first statement is the isoperimetric theorem, the second a celebrated conjecture of Lord Rayleigh. Our tables yield sound inductive evidence for both statements but, of course, no proof.

The situation has changed since we considered these tables in sect. 10.1 and 10.4. In the meantime we have seen a proof for the isoperimetric theorem (sect. 10.6–10.8, ex. 10.1–10.15). The geometrical minimum property of the circle, inductively supported by Table II, has been proved. It is natural to expect that the analogous physical minimum property of the circle, inductively supported by Table III, will also turn out to be true. In expecting this we follow an important pattern of plausible inference:

$$\frac{\begin{array}{c} A \text{ analogous to } B \\ B \text{ true} \end{array}}{A \text{ more credible}}$$

A conjecture becomes more credible when an analogous conjecture turns out to be true.

The application of this pattern to the situation discussed seems sensible. Yet there are further promising indications in this situation.

5. Deepening the analogy. The Tables II and III, side by side, seem to offer further suggestions. The ten figures considered do not appear in exactly the same sequence in both tables. There is something peculiar about this sequence. The arrangement in Table II appears not very different from that in Table III, but this is not the main point. The tables contain various kinds of figures: rectangles, triangles, sectors. How are the *figures of the same kind* arranged? How would a shorter table look listing only figures of one kind? The tables contain a few regular figures: the equilateral triangle, the square, and, let us not forget it, the circle. How are the regular figures arranged? Could we compare somehow figures of different kinds, for instance, triangles and sectors? Could we broaden the inductive basis by adding further figures to our tables? (In this we are much restricted. It is not difficult to compute areas and perimeters, but the principal frequency is difficult to handle and its explicit expression is known in very few cases only.) Eventually we obtain Table IV.

Table IV exhibits a remarkable parallelism between these two quantities depending on the shape of a variable plane figure: the perimeter and the principal frequency. (We should not forget that the area of the variable figure is fixed, = 1.) If we know the perimeter, we are by no means able to compute the principal frequency or vice versa. Yet, judging from Table IV, we should think that, in many simple cases, these two quantities *vary in the same direction*. Consider the two columns of numerical data in this table and pass from any row to the next row: if there is an increase in one of the columns, there is a corresponding increase in the other, and if there is a decrease in one of the columns, there is a corresponding decrease in the other.

Table IV

Perimeters and principal frequencies of figures of equal area

Figure	Perimeter	Pr. frequency
Rectangles:		
1 : 1 (square)	4.00	4.443
3 : 2	4.08	4.624
2 : 1	4.24	4.967
3 : 1	4.64	5.736
Triangles:		
60° 60° 60°	4.56	4.774
45° 45° 90°	4.84	4.967
30° 60° 90°	5.08	5.157
Sectors:		
180° (semicircle)	4.10	4.803
90° (quadrant)	4.03	4.551
60° (sextant)	4.21	4.616
45°	4.44	4.755
36°	4.68	4.916
30°	4.93	5.084
Regular figures:		
circle	3.55	4.261
square	4.00	4.443
equilateral triangle	4.56	4.774
Triangles versus sectors:		
tr. 60° 60° 60°	4.56	4.774
sector 60°	4.21	4.616
tr. 45° 45° 90°	4.84	4.967
sector 45°	4.44	4.755
tr. 30° 60° 90°	5.08	5.157
sector 30°	4.93	5.084

Let us focus our attention on the rectangles. If the ratio of the length to the width increases from 1 to ∞, so that the shape varies from a square to an infinitely elongated rectangle, both the perimeter and the principal frequency seem to increase steadily. The square which, being a regular figure, is "nearest" to the circle among all quadrilaterals, has the minimum perimeter and also the minimum principal frequency. Of the three triangles listed, the equilateral triangle which, being a regular figure, is "nearest" to the circle among all triangles has the minimum perimeter and also the minimum principal frequency. The behavior of the sectors is more complex. As the angle of the sector varies from 180° to 0°, the perimeter first decreases, attains a minimum, and then increases; and the principal frequency varies in the same manner. Let us now look at the regular figures. The equilateral triangle has 3 axes of symmetry, the square has 4 such axes, and the circle an infinity.

As far as we can see from Table IV, both the perimeter and the principal frequency seem to decrease as the number of the axes of symmetry increases. In the last section of Table IV we matched each triangle against the sector whose angle is equal to the least angle of the triangle. In all three cases, the sector turned out to be "more circular," having the shorter perimeter and the lower principal frequency.

What we definitely know about these regularities goes, of course, only as far as Table IV goes. That these regularities hold beyond the limits of the experimental material collected is suggested and rendered plausible by Table IV, but is by no means proved. And so Table IV led us to several new conjectures which are similar to Rayleigh's conjecture although, of course, of much more limited scope.

How does Table IV influence our confidence in Rayleigh's conjecture? Can we find in Table IV any reasonable ground for it that we did not notice before in discussing the Tables II and III?

We certainly can. First of all, the Table IV contains a few more particular cases in which Rayleigh's conjecture is verified (the 30° 60° 90° triangle, the sectors with opening 45°, 36°, and 30°). Yet there is more than that. The analogy between the isoperimetric theorem and Rayleigh's conjecture has been considerably deepened; the facts listed in Table IV add several new aspects to this analogy. Now it seems to be reasonable to consider a conclusion from analogy as becoming stronger if the analogy itself, on which the conclusion is based, becomes stronger. And so Table IV considerably strengthens Rayleigh's case.

6. Shaded analogical inference. Yet there is still something more. As we have observed, Table IV suggests several conjectures which are analogous to (but of more limited scope than) Rayleigh's conjecture. Table IV suggests these conjectures and lends them some plausibility too. Yet this circumstance quite reasonably raises somewhat the plausibility of Rayleigh's original conjecture. If you think so too, you think according to the following pattern:

$$\begin{array}{c} A \text{ analogous to } B \\ B \text{ more credible} \\ \hline A \text{ somewhat more credible} \end{array}$$

A conjecture becomes somewhat more credible when an analogous conjecture becomes more credible. This is a weakened or *shaded* form of the pattern formulated in sect. 4.

EXAMPLES AND COMMENTS ON CHAPTER XII

1. Table I, exhibiting some inductive evidence for Euler's conjecture mentioned in sect. 1, is very similar to the table in sect. 1.3, or to Tables I,

II, and III in ch. IV, or to Euler's table given in support of his "Most Extraordinary Law of the Numbers Concerning the Sum of Their Divisors," see sect. 6.2. These tables resemble also two tables given in ch. III, one in sect. 3.1 (polyhedra), the other in sect. 3.12 (partitions of space). To which one of these two is the resemblance closer?

2. Euler, having verified his "Most Extraordinary Law" (cf. sect. 6.2) for $n = 1, 2, 3, 4, \ldots 20$, proceeds to verify it for $n = 101$ and $n = 301$. He had a good reason to examine 101 and 301 rather than 21 and 22 (which he clearly states at the outset of No. 7 of his memoir). Ignoring, or remembering only vaguely, the contents of Euler's Law, would you think that the verification of his two cases (101 and 301) has more probative value than would have the verification of the two next cases (21 and 22)?

3. Of a triangle, let a, b, and c denote the sides, $2p = a + b + c$ the perimeter, A the area.

Check Heron's formula

$$A^2 = p(p - a)(p - b)(p - c)$$

in as many ways as you can.

4. We consider a quadrilateral inscribed in a circle. Let a, b, c, and d denote the sides, $2p = a + b + c + d$ the perimeter, A the area.

It is asserted that

$$A^2 = (p - a)(p - b)(p - c)(p - d).$$

Check this assertion in as many ways as you can. Have you any comment?

5. Let V denote the volume of a tetrahedron and

$$\begin{matrix} a, & b, & c, \\ e, & f, & g \end{matrix}$$

the lengths of its six edges; the edges a, b, and c end in the same vertex of the tetrahedron, e is the edge opposite to a, f to b, and g to c. (Two edges of a tetrahedron are called opposite to each other if they have no vertex in common.) The edges e, f, and g are the three sides of a face of the tetrahedron, opposite to the vertex in which a, b, and c end. It is asserted that

$$\begin{aligned} 144V^2 = {} & 4a^2b^2c^2 + (b^2 + c^2 - e^2)(c^2 + a^2 - f^2)(a^2 + b^2 - g^2) \\ & - a^2(b^2 + c^2 - e^2)^2 - b^2(c^2 + a^2 - f^2)^2 - c^2(a^2 + b^2 - g^2)^2. \end{aligned}$$

Check this assertion in as many ways as you can. [Is the expression proposed for V symmetric in the six edges?]

6. Set

$$a^n + b^n + c^n = s_n$$

for $n = 1, 2, 3, \ldots$ and define p, q, and r by the identity in x

$$(x - a)(x - b)(x - c) = x^3 - px^2 + qx - r$$

so that

$$p = a + b + c,$$

$$q = ab + ac + bc,$$

$$r = abc.$$

(In the usual terminology, p, q, and r are the "elementary symmetric functions" of a, b, and c, and s_n a "sum of like powers.") Obviously, $p = s_1$. It is asserted that, for arbitrary values of a, b, and c,

$$q = \frac{2s_1^5 - 5s_1^2 s_3 + 3s_5}{5(s_1^3 - s_3)},$$

$$r = \frac{s_1^6 - 5s_1^3 s_3 - 5s_3^2 + 9s_1 s_5}{15(s_1^3 - s_3)}$$

provided that the denominator does not vanish. Check these formulas in the particular case $a = 1$, $b = 2$, $c = 3$ and in three more cases displayed in the table:

Case	a	b	c
(1)	1	2	3
(2)	1	2	−3
(3)	1	2	0
(4)	1	2	−2

Devise further checks. Especially, try to generalize the cases (2), (3), and (4).

7. Let A, B_1, B_2, B_3, and B_4 have the meaning given them in sect. 2. Does the verification of B_4, coming after that of B_1, B_2, and B_3, supply additional inductive evidence for A?

8. Let us recall Euler's "Most Extraordinary Law" and the meaning of the abbreviations T, C_1, C_2, C_3, $\ldots$ C_1^*, C_2^*, C_3^*, $\ldots$ explained in sect. 6.3. Euler supported the theorem T, when he was not yet able to prove it, inductively, by verifying its consequences C_1, C_2, C_3, $\ldots$ C_{20}. (He went even further, perhaps.) Then he discovered that also C_1^*, C_2^*, C_3^*, $\ldots$ are consequences of T, and verified C_1^*, C_2^*, $\ldots$ C_{20}^*, C_{101}^*, C_{301}^*. Thanks to these new verifications Euler's confidence was, presumably, much strengthened: but was it justifiably strengthened? [Closer attention to detail is needed here than in ex. 2.]

9. We return to Euler's conjecture discussed in sect. 1; for the sake of brevity, we call it the "conjecture E." Let us note concisely the meaning of this abbreviation,

$$E: \qquad 8n + 3 = x^2 + 2p.$$

The idea that led Euler to his conjecture E deserves mention. Euler devoted much of his work to those celebrated propositions of Number Theory that Fermat has stated without proof. One of these (we call it the "conjecture F") says that any integer is the sum of three trigonal numbers. Let us note concisely the meaning of this abbreviation,

$$F: \qquad n = \frac{x(x-1)}{2} + \frac{y(y-1)}{2} + \frac{z(z-1)}{2}.$$

Euler observed that *if* his conjecture E were true, Fermat's conjecture F would easily follow. That is, Euler satisfied himself that E implies F. (For details, see the next ex. 10.) Bent on proving Fermat's conjecture F, Euler naturally desired that his conjecture E should be true. Is this mere wishful thinking? I do not think so; the relations considered yield some weak but not unreasonable ground for believing Euler's conjecture E according to the following scheme:

E implies F
F credible

E somewhat more credible

Here is another pattern of plausible inference. The reader should compare it with the fundamental inductive pattern.

10. In proving that E implies F (in the notation of the foregoing ex. 9), Euler used a deeper result which he proved previously: a prime number of the form $4n + 1$ is a sum of two squares. (This was discussed inductively in ex. 4.4.) Taking this for granted, prove that actually E implies F.

11. After having conceived his conjecture discussed in sect. 3, Euler tested it by computing the first zeros of his series for a few values of n. (By a "first zero" we mean a zero the absolute value of which is a minimum. If x is a first zero of the series in question, also $-x$ is a zero, and x and $-x$ are "first zeros." Therefore, x is real if, and only if, x^2 is positive.) Of course, Euler had to compute these zeros approximately. A method (Daniel Bernoulli's method) which he frequently used for such a purpose yielded the following sequences of approximate values for the first zero x in the cases $n = 1/2, 1/3, 1/4$.

$n = 1/2$	$n = 1/3$	$n = 1/4$
$4x^2 \sim 3.000$	$9x^2 \sim 4.0000$	$16x^2 \sim 5.0000$
3.281	4.2424	5.2232
3.291	4.2528	5.2302
3.304	4.2532	5.2304

In all three cases, the approximate values seem to tend to a positive limit regularly and rather rapidly. Euler takes this as a sign that the first zeros are real and sees herein a confirmation of his conjecture.

Let us realize the general scheme of Euler's heuristic conclusion. Let A stand for his conjecture explained in sect. 3, concerning the reality of the zeros of his series. Let B stand for the fact that for $n = 1/2$ the first zero is real. Obviously A implies B. Now Euler did not prove B, he only made B more credible. Therefore we have here the following pattern of plausible inference

$$\frac{\begin{array}{c} A \text{ implies } B \\ B \text{ more credible} \end{array}}{A \text{ somewhat more credible}}$$

The second premise is weaker than the second premise of the fundamental inductive pattern. The word "somewhat" is inserted to emphasize that also the conclusion is weaker than in the fundamental inductive pattern.

12. A modern mathematician can derive a more stringent heuristic conclusion from the numerical data of the foregoing ex. 11 than Euler himself derived. It can be shown that if Euler's series has only real zeros, the successive approximate values obtained by Daniel Bernoulli's method form necessarily an *increasing* sequence.[7] Let A stand for the same conjecture as in the foregoing ex. 11, but let B denote now another statement, namely the following: "For $n = 1/2$, the first four approximations obtained by Daniel Bernoulli's method form an increasing sequence, and the same holds for $n = 1/3$ and $n = 1/4$." Then both premises of the fundamental inductive pattern are known to hold:

$$\begin{array}{c} A \text{ implies } B \\ B \text{ true} \end{array}$$

and the resulting heuristic conclusion is stronger.

Two remarks may be added to the foregoing.

(1) Euler did not formulate the property just quoted of Daniel Bernoulli's method and certainly did not prove it. Yet there is a good chance that, in the possession of a vast experience with this method, he had some sort of inductive knowledge of it. So Euler, although he did not draw the modern mathematician's inference explicitly, possessed it in a less clarified form. And, presumably, he had in his rich mathematical background still other indications which he could not quite formulate and which we could not yet clarify today.

---

[7] See the author's paper "Remarks on power series," *Acta Scientiarum Mathematicarum*, v. 12B, 1950, p. 199–203.

(2) The numerical data quoted in ex. 11 led the author to suspect the general theorem proved l.c. This is a small but concrete example of the use of the inductive method in mathematical research.

13. In ch. IV we investigated inductively the sum of four odd squares; see sect. 4.3–4.6, Table I. Later we tackled the analogous problems involving four arbitrary squares and eight squares; see ex. 4.10–4.23 and Tables II and III. The former investigation certainly helped us to recognize the law in the latter cases. Should our confidence in the result of the latter investigation also be enhanced by the result of the former?

14. *Inductive conclusion from fruitless efforts.* Construct, by ruler and compasses, the side of a square equal in area to a circle of given radius. This is the strict formulation of the famous problem of the quadrature of the circle, conceived by the Greeks. It was not forgotten in the Middle Ages, although we cannot believe that many people then understood its strict formulation; Dante refers to it at the theological culmination of the *Divina Commedia*, toward the end of the concluding Canto. The problem was about two thousand years old as the French Academy resolved that manuscripts purporting to square the circle will not be examined. Was the Academy narrow-minded? I do not think so; after the fruitless efforts of thousands of people in thousands of years there was some ground to suspect that the problem is insoluble.

You are moved to give up a task that withstands your repeated efforts. You desist only after many and great efforts if you are stubborn or deeply concerned. You desist after a few cursory trials if you are easy going or not seriously concerned. Yet in any case there is a sort of inductive conclusion. The conjecture under consideration is:

A. It is impossible to do this task.

You observe:

B. Even I cannot do this task.

This, in itself, is *very* unlikely indeed. Yet certainly

$$A \text{ implies } B$$

and so your observation of B renders A more credible, by the fundamental inductive pattern.

The impossibility of squaring the circle, strictly formulated, was proved in 1882, by Lindemann, after the basic work of Hermite. There are other similar problems dating from the Greeks (the Trisection of an Angle and the Duplication of the Cube) that, after the accumulated evidence of fruitless efforts, have been ultimately proved insoluble. After fruitless efforts to construct a "perpetuum mobile" the physicists formulated the "principle of the impossibility of a perpetual motion," and this principle turned out remarkably fruitful.

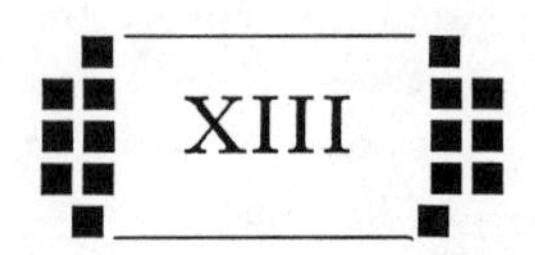

XIII

FURTHER PATTERNS AND FIRST LINKS

When we have intuitively understood some simple propositions . . . it is useful to go through them with a continuous, uninterrupted motion of thought, to meditate upon their mutual relations, and to conceive distinctly several of them, as many as possible, simultaneously. In this manner our knowledge will grow more certain, and the capacity of the mind will notably increase.—DESCARTES[1]

1. Examining a consequence. We consider a situation which frequently occurs in mathematical research. We wish to decide whether a clearly formulated mathematical proposition A is true or not. We have, perhaps, some intuitive confidence in the truth of A, but that is not enough: we wish to prove A or disprove it. We work at this problem, but without decisive success. After a while we notice a consequence B of A. This B is a clearly formulated mathematical proposition of which we know that it follows from A:

$$A \text{ implies } B.$$

Yet we do not know whether B is true or not. Now it seems that B is more accessible than A; for some reason or other we have the impression that we shall have better success with B than we had with A. Therefore, we switch to examining B. We work to answer the question: is B true or false? Finally we succeed in answering it. *How does this answer influence our confidence in A?*

That depends on the answer. If we find that B, this consequence of A, is false, we can infer with certainty that A must also be false. Yet if we find that B is true, there is no demonstrative inference: although its consequence B turned out to be true, A itself could be false. Yet there is a heuristic inference: since its consequence B turned out to be true, A itself seems to

[1] The eleventh of his Rules for the Direction of the Mind. See *Oeuvres*, edited by Adam and Tannery, vol. 10, 1908, p. 407.

deserve more confidence. According to the nature of our result concerning B, we follow a demonstrative or a heuristic pattern:

Demonstrative	*Heuristic*
A implies B	A implies B
B false	B true
A false	A more credible

We met these patterns already in sect. 12.1 where we called the heuristic pattern the fundamental inductive pattern. We shall meet with similar but different patterns in the following sections.

2. Examining a possible ground. We consider another situation that frequently occurs in mathematical research. We wish to decide whether the clearly formulated proposition A is true or not, we wish to prove A or disprove it. After some indecisive work we hit upon another proposition B from which A would follow. We do not know whether B is true or not, but we have satisfied ourselves that

$$A \text{ is implied by } B.$$

Thus, if we could prove B, the desired A would follow; B is a possible ground for A. We may be tired of A, or B may appear to us more promising than A; for some reason or other we switch to examining B. Our aim is now to prove or disprove B. Finally we succeed. How will our result concerning B influence our confidence in A?

That depends on the nature of our result. If we find that B is true, we can conclude that A which is implied by B (follows from B, is a consequence of B) is also true. Yet if we find that B is false, there is no demonstrative conclusion: A could still be true. But we have been obliged to discard a possible ground for A, we have one chance less to prove A, our hope to prove A from B has been wrecked: if there is any change at all in our confidence in A in consequence of the disproof of B, it can only be a change for the worse. In short, according to the nature of our result concerning B, we follow a demonstrative or a heuristic pattern:

Demonstrative	*Heuristic*
A implied by B	A implied by B
B true	B false
A true	A less credible

Observe that the first premise is the same in both patterns. The second premises are diametrically opposite, and the conclusions are also opposite, although not quite as far apart.

The demonstrative inference follows a classical pattern, the "modus ponens" of the so-called hypothetical syllogism. The heuristic pattern is similar to, but different from, the fundamental inductive pattern, see sect. 1 or sect. 12.1. We can state the heuristic inference in words: *our confidence in a conjecture can only diminish when a possible ground for the conjecture is exploded.*

3. Examining a conflicting conjecture. We consider a situation which is not too usual in mathematical research but frequently occurs in the natural sciences. We examine two conflicting, incompatible conjectures A and B. When we say that A conflicts with B or

A is incompatible with B

we mean that the truth of one of the two propositions A and B necessarily implies the falsity of the other. Thus, A may be true or not and B may be true or not; we do not know which is the case except that we know that both cannot be true. They could both be false, however. A naturalist proposed the conjecture A to explain some phenomenon, another naturalist proposed the conjecture B to explain the same phenomenon differently. The explanations are incompatible; both naturalists cannot be right, but both could be wrong.

If one of the conjectures, say B, has been proved to be right, then the fate of the other is also definitely decided: A must be wrong. If, however, B has been disproved, the fate of A is not yet definitely settled: also A could be wrong. Yet, undeniably, by the disproof of a rival conjecture incompatible with it, A can only gain. (The naturalist who invented A would certainly think so.) And so we have again two patterns:

Demonstrative	*Heuristic*
A incompatible with B	A incompatible with B
B true	B false
A false	A more credible

Our confidence in a conjecture can only increase when an incompatible rival conjecture is exploded.

4. Logical terms. In the foregoing three sections we have seen three pairs of patterns. Each pair consists of a demonstrative pattern and of a heuristic pattern; the three demonstrative patterns are related to each other and the corresponding three heuristic patterns seem to be correspondingly related. The relations between the demonstrative patterns are clear relations of formal logic. In the next section we shall try to clarify the relations between the heuristic patterns. The present section prepares us for the next section by discussing a few simple terms of formal logic.[2]

[2] We treat here formal logic in the "old-fashioned" manner (using ordinary language and avoiding logical symbols as much as feasible) but with a few modern ideas. Some of the simpler logical symbols will be used incidentally later, especially in sect. 7.

(1) The term *proposition* may be taken in a more general meaning, but most of the time it will be sufficient and even advantageous to think of some clearly formulated mathematical proposition of which *for the moment we do not know whether it is true or not.* (A good example of a proposition for a more advanced reader is the celebrated "Riemann hypothesis": Riemann's ξ-function has only real zeros. We do not know, in spite of the efforts of many excellent mathematicians, whether this proposition is true or false.) We shall use capitals A, B, C, . . . to denote propositions.

(2) The *negation* of the proposition A is a proposition that is true if, and only if, A is false. We let non-A stand for the negation of A.

(3) The two statements "A is false" and "non-A is true" amount exactly to the same. We can substitute one for the other in any context without changing the import, the truth, or the falsity, of the whole text. Two statements which can be so substituted for each other are termed *equivalent.* Thus, the statement "A is false" is equivalent to the statement "non-A is true." It will be convenient to write this in the abbreviated form:

"A false" eq. "non-A true."

It is also correct to say that

"A true" eq. "non-A false."

"non-A true" eq. "A false,"

"non-A false" eq. "A true."

(4) We say that the two propositions A and B are *incompatible* with each other, if both cannot be true. The proposition A can be true or false, B can be true or false; if we consider A and B jointly, four different cases can arise:

A true, B true	A true, B false
A false, B true	A false, B false.

If we say that A is incompatible with B, we mean that the first of these four cases (in the north-west corner) is excluded. Incompatibility is always mutual. Therefore,

"A incompatible with B" eq. "B incompatible with A."

(5) We say that A implies B (or B is implied by A, or B follows from A, or B is a consequence of A, etc.) if A and non-B are incompatible. Thus the concept of implication is characterized by the following equivalence:

"A implies B" eq. "A incompatible with non-B."

To know that A implies B is important. For the moment we do not know whether A is true or not and we are in the same state of ignorance concerning B. If, however, it should turn out some day that A is true, we shall know right away that non-B must be false and so B must be true.

We know that

"A incompatible with non-B" eq. "non-B incompatible with A."

We know also that

"non-B incompatible with A" eq. "non-B implies non-A."

From the chain of the last three equivalences we conclude:

"A implies B" eq. "non-B implies non-A."

This last equivalence is quite important in itself and it will be essential in the following consideration.

(6) The few points of formal logic discussed in this section enable us already to clarify the relation between the demonstrative patterns encountered in the three foregoing sections.

Let us start from the demonstrative pattern formulated in sect. 1 (the "modus tollens"):

A implies B
B false

A false

It is understood that this pattern is generally applicable. Let us apply it in substituting non-A for A and non-B for B. We obtain

non-A implies non-B
non-B false
---
non-A false

We have seen, however, in the foregoing that

"non-A implies non-B" eq. "B implies A"

"non-B false" eq. "B true"

"non-A false" eq. "A true"

Let us substitute for the premises and the conclusion of the last considered pattern the three corresponding equivalent statements here displayed. Then we obtain:

B implies A
B true

A true

which is the demonstrative pattern of sect. 2, the "modus ponens."

We leave to the reader to derive similarly the demonstrative pattern of sect. 3 from that of sect. 1.

5. Logical links between patterns of plausible inference. The discussion in the foregoing section was merely preparatory. We did not discuss those classical points of demonstrative logic for their own sake, but in order to prepare ourselves for the study of plausible inference. We considered the derivation of the "modus ponens" from the "modus tollens" not in some vain hope of presenting something new and surprising concerning such classical matters, but as a preparation for such questions as the following: Can we derive by pure formal logic the heuristic pattern of sect. 2 from the heuristic pattern of sect. 1?

No, obviously, we cannot. In fact, these patterns contain such statements as "*A* becomes more credible" or "*A* becomes less credible." Although everybody understands what this means, the consistent formal logician refuses to understand such statements, and he is even right. Pure formal logic has no place for such statements; it has no way to handle them.

We could, however, widen the domain of formal logic in an appropriate way. The main point seems to be to add the following equivalence to the classical contents of formal logic: "non-*A* becomes more credible" is equivalent to "*A* becomes less credible." In abbreviation

"non-*A* more credible" eq. "*A* less credible."

In admitting this, we obtain a useful tool. Now we can adapt the procedure of sect. 4 (6) to our present aim and proceed as follows. We start from the fundamental inductive pattern, given in sect. 1. We apply it to non-*A* and non-*B*, instead of applying it to *A* and *B*; that is, we substitute non-*A* for *A* and non-*B* for *B* in it. Then we apply three equivalences; two of these are purely logical (and discussed, by the way, in the foregoing sect. 4) and the third is the essential new equivalence introduced in the present section. By these steps we finally arrive at the heuristic pattern given in sect. 2.

It may be left to the reader to write down this derivation in detail and to derive the heuristic pattern of sect. 3 in the same manner from the fundamental inductive pattern of sect. 1.

6. Shaded inference. It seems to me that inductive reasoning in the mathematical domain is easier to study than in the physical domain. The reason is simple enough. In asking a mathematical question, you may hope to obtain a completely unambiguous answer, a perfectly sharp Yes or No. In addressing a question to Nature, you cannot hope to obtain an answer without some margin of uncertainty. You predict that a lunar eclipse will begin (the shadow will indent the disk of the moon) at such and such a time. Actually, you observe the beginning of the eclipse 4 minutes later than predicted. According to the standards of Greek astronomy your prediction would be amazingly correct, according to modern standards it is scandalously incorrect. A given discrepancy between prediction and observation can be interpreted as confirmation or refutation. Such interpretation depends on some kind of plausible reasoning the difficulties of which in "physical situations" begin a step earlier than in "mathematical situations." We shall try to reduce this distinction to its simplest expression.

Suppose that we are investigating a mathematical conjecture by examining its consequences. Let A stand for the conjecture and B for one of its consequences, so that A implies B. We arrive at a final decision concerning B: we disprove B or we prove it and, accordingly, we face one or the other of the following two situations:

A implies B	A implies B
B false	B true
———————	———————

We shall call these "mathematical situations." We have considered them repeatedly and we know what reasonable inference we can draw from each.

Suppose now that we are investigating a physical conjecture A and that we test experimentally one of its consequences B. We cannot arrive at an absolute decision concerning B; our experiments may show, however, that B, or its contrary, is very hard to believe. Accordingly, we face one or the other of the following two situations:

A implies B	A implies B
B scarcely credible	B almost certain
———————	———————

We call these "physical situations." What inference is reasonable in these situations? (The empty space under the horizontal line that suggests the word "therefore" symbolizes the open question.)

In each of the four situations considered we have two data or *premises*. The first premise is the same in all four situations; all the difference between them hinges on the second premise. This second premise is on the level of pure formal logic in the "mathematical" situations, but on a much vaguer level in the "physical" situations. This difference seems to me essential; it may account for the additional difficulties of the physical situations.

Let us survey the four situations "with a continuous uninterrupted motion of thought," as Descartes liked to say (see the motto at the beginning of this chapter). Let us imagine that our confidence in B changes gradually, varies "continuously." We imagine that B becomes less credible, then still less credible, scarcely believable, and finally false. On the other hand, we imagine that B becomes more credible, then still more credible, practically certain, and finally true. If the *strength of our conclusion varies continually in the same direction as the strength of our confidence in* B, there is little doubt what our conclusion should be since the two extreme cases (B false, B true) are clear. In this manner we arrive at the following patterns:

$$\frac{\begin{array}{c} A \text{ implies } B \\ B \text{ less credible} \end{array}}{A \text{ less credible}} \qquad \frac{\begin{array}{c} A \text{ implies } B \\ B \text{ more credible} \end{array}}{A \text{ somewhat more credible}}$$

The word "somewhat" in the second pattern is inserted to remind us that the conclusion is, of course, weaker than in the fundamental inductive pattern. *Our confidence in a conjecture is influenced by our confidence in one of its consequences and varies in the same direction.* We shall call these patterns *shaded*; the first is a shaded demonstrative pattern, the second is the shaded version of the fundamental inductive pattern. The term "shaded" intends to indicate the weakening of the second premise: "less credible" instead of "false"; "more credible" instead of "true." We have already used this term in this meaning in sect. 12.6.

We obtained the shaded patterns just introduced from their extreme cases, the "modus tollens" and the fundamental inductive pattern discussed in sect. 1, by weakening the second premise. In the same manner we can obtain other shaded patterns from the patterns formulated in sects. 2 and 3. We state just one here (all are listed in the next section). The heuristic pattern of sect. 3 yields the following shaded pattern:

$$\frac{\begin{array}{c} A \text{ incompatible with } B \\ B \text{ less credible} \end{array}}{A \text{ somewhat more credible}}$$

7. A table. In order to list concisely the patterns discussed in this chapter, it will be convenient to use some abbreviations. We write

$A \rightarrow B$ for A implies B,

$A \leftarrow B$ for A is implied by B,

$A \mid B$ for A incompatible with B.

The symbols introduced are used by some authors writing on symbolic logic.[3] In this notation, the two formulas

$$A \rightarrow B, \qquad B \leftarrow A$$

are exactly equivalent and so are the formulas

$$A \mid B, \qquad B \mid A.$$

[3] D. Hilbert and W. Ackerman, *Grundzüge der theoretischen Logik.*

We shall also abbreviate "credible" as "cr." and "somewhat" as "s." See Table I.

Table I

	(1) Demonstrative	(2) Shaded Demonstrative	(3) Shaded Inductive	(4) Inductive
1. Examining a consequence	$A \rightarrow B$ B false ——— A false	$A \rightarrow B$ B less cr. ——— A less cr.	$A \rightarrow B$ B more cr. ——— A s. more cr.	$A \rightarrow B$ B true ——— A more cr.
2. Examining a possible ground	$A \leftarrow B$ B true ——— A true	$A \leftarrow B$ B more cr. ——— A more cr.	$A \leftarrow B$ B less cr. ——— A s. less cr.	$A \leftarrow B$ B false ——— A less cr.
3. Examining a conflicting conjecture	$A \mid B$ B true ——— A false	$A \mid B$ B more cr. ——— A less cr.	$A \mid B$ B less cr. ——— A s. more cr.	$A \mid B$ B false ——— A more cr.

8. Combination of simple patterns. The following situation can easily arise in mathematical research. We investigate a theorem A. This theorem A is clearly formulated, but we do not know and we wish to find out whether it is true or false. After a while we hit upon a possible ground: we see that A can be derived from another theorem H

A is implied by H

and so we try to prove H. We do not succeed in proving H, but we notice that one of its consequences B is true. The situation is:

A implied by H
B implied by H
B true
———————

Is there a reasonable inference concerning A from these premises?

There is one, I think, and we can even obtain it by combining two of the patterns surveyed in the foregoing section. In fact, the fundamental inductive pattern yields:

H implies B
B true
———————
H more credible

In obtaining this conclusion, we have used only two of our three premises. Let us combine the unused third premise with the conclusion just obtained;

one of our shaded patterns (in the intersection of the second row with the second column of Table I in sect. 7) yields:

$$\frac{\begin{array}{c} A \text{ implied by } H \\ H \text{ more credible} \end{array}}{A \text{ more credible}}$$

The result is (as it should be) pretty obvious in itself: from the verification of the consequence B of its possible ground H some credit is reflected upon the proposition A itself.

9. On inference from analogy. The situation discussed in the foregoing section can be construed as a link between the patterns discussed in this chapter and one of the most conspicuous forms of plausible inference: the conclusion from analogy.

I do not think that it is possible to explain the idea of analogy in completely definite terms of formal logic; at any rate, I have no ambition to explain it so. As we have discussed before, in sect. 2.4, analogy has to do with similarity *and* the intentions of the thinker. If you notice some similarity between two objects (or, preferably, between two systems of objects) and intend to reduce this similarity to definite concepts, you think analogically.

For example, you notice some similarity between two theorems A and B; you observe some common points. Perhaps, you think, it will be possible some day to imagine a more comprehensive theorem H that brings out all the essential common points and from which both A and B would naturally follow. If you think so you begin to think analogically.[4]

At any rate, let us consider the analogy of two theorems A and B as the intention to discover a common ground H from which both A and B would follow:

$$A \text{ implied by } H, \qquad B \text{ implied by } H.$$

Let us not forget that we do not *have* H; we just *hope* that there is such an H.

Now we succeed in proving one of the two analogous theorems, say B. How does this event influence our confidence in the other theorem A? The situation has something in common with the situation considered in the foregoing sect. 8. There we reached a reasonable conclusion expressed by the compound pattern

$$\frac{\begin{array}{c} A \text{ implied by } H \\ B \text{ implied by } H \\ B \text{ true} \end{array}}{A \text{ more credible}}$$

[4] Thus, the isoperimetric theorem and Rayleigh's conjecture, compared with each other in sect. 12.4, may suggest the idea of a common generalization.

There is, of course, the important difference that now we do not have H, we just hope for H. With this proviso, however, we can regard the two premises

$$\begin{array}{c} A \text{ implied by } H \\ B \text{ implied by } H \end{array}$$

as equivalent to one:

$$A \text{ analogous to } B.$$

In substituting this one premise for those two premises in the above compound pattern we arrive at the fundamental pattern of plausible inference first exhibited in sect. 12.4:

$$\begin{array}{c} A \text{ analogous to } B \\ B \text{ true} \\ \hline A \text{ more credible} \end{array}$$

10. Qualified inference. We come back again to the fundamental inductive pattern. It is the first pattern that we introduced and it is the most conspicuous form of plausible reasoning. It is concerned with the verification of a consequence of a conjecture and the resulting change in our opinion. It says something about the direction of this change; such a verification can only increase our confidence in the conjecture. It says nothing about the amount of the change; the increase of confidence can be great or small. Indeed, it can be tremendously great or ridiculously small.

The aim of the present section is to clarify the circumstances on which such an important difference depends. We begin by recalling one of our examples (sect. 12.3).

A defendant is accused of having blown up the yacht of his girl friend's father and the prosecution produces a receipt signed by the defendant acknowledging the purchase of such and such an amount of dynamite.

The evidence against the defendant appears very strong. Why does it appear so? Let us insist on the general features of the case. Two statements play an essential role.

A: the defendant blew up that yacht.

B: the defendant acquired explosives.

At the beginning of the proceedings, the court has to consider A as a conjecture. The prosecution works to render A more credible to the jurors, the defense works to render it less credible.

At the beginning of the proceedings also B has to be considered as a conjecture. Later, after the introduction of that receipt in court (the authenticity of the signature was not challenged by the defense) B has to be considered as a proven fact.

Certain relations between A and B, however, should be clear from the beginning.

A without B is impossible. If the defendant blew up the yacht, he had some explosives. He had to acquire these explosives somehow: by purchase, stealing, gift, inheritance or otherwise. That is

A implies B.

B without A is not impossible, but must appear extremely unlikely from the outset. To buy dynamite is very unusual anyhow for an ordinary citizen. To buy dynamite without the intention of blowing up something or somebody would be nonsense. It was easy to suspect that the defendant had quite strong emotional and financial grounds for blowing up that yacht. It was difficult to suspect any purpose for the purchase of dynamite, except blowing up the yacht. And so, as we said, B without A appears extremely unlikely.

Let us nail down this important constituent of the situation: *the credibility of B, before the event, viewed under the assumption that A is not true.* We shall abbreviate this precise but long description as "the credibility of B without A." Thus, we can say:

B without A is hardly credible.

Now, we can see the essential premises and the whole pattern of the plausible inference that impressed us with its cogency:

A implies B
B without A hardly credible
B true
———————————————
A very much more credible

For better understanding let us imagine that that important constituent of the situation, the credibility of B without A, changes gradually, varies continuously between its extreme cases.

A implies B. If, conversely, also B implies A, so that A and B imply each other mutually, the credibility of B without A attains its minimum, is nil. In this case, if B is true, also A is true, so that the credibility of A attains its maximum.

A implies B. That is, B is certain when A is true. If the credibility of B without A approaches its maximum, B is almost certain when A is false. Therefore, B is almost certain anyway. When an event happens that looks almost certain in advance, we do not get much new information and so we cannot draw surprising consequences. (The purchase of a loaf of bread, for instance, can hardly ever yield such a strong circumstantial evidence, as the purchase of dynamite.)

Let us assume, as in sect. 6, that the strength of the conclusion varies *continually in the same direction* when that influential factor, the credibility of B without A, changes without change of direction. Then these two must vary in opposite directions and we arrive so at an important qualification of the strength of the conclusion in the fundamental inductive pattern:

$$\frac{\begin{array}{c} A \text{ implies } B \\ B \text{ true} \end{array}}{A \text{ more credible}}$$

The strength of the conclusion increases when the credibility of B without A decreases.

Let us put side by side the two extreme cases:

$$\frac{\begin{array}{c} \left\{\begin{array}{l} A \text{ implies } B \\ B \text{ without } A \text{ hardly credible} \end{array}\right. \\ B \text{ true} \end{array}}{A \text{ very much more credible}} \qquad \frac{\begin{array}{c} \left\{\begin{array}{l} A \text{ implies } B \\ B \text{ almost certain anyway} \end{array}\right. \\ B \text{ true} \end{array}}{A \text{ very little more credible}}$$

The first two premises are bracketed to express that the second is to be considered as a qualification of the first. That A implies B is the first premise in the fundamental inductive pattern. Here we qualify this premise; we add a modification which counts heavily in determining the strength of the conclusion. For the sake of comparison let us remember that in sect. 6 we modified the fundamental inductive pattern in another direction, in weakening its second premise.

11. On successive verifications. We have verified already n consequences $B_1, B_2, \ldots, B_n$ of a certain conjecture A. Now we proceed to a next consequence B_{n+1}, we test it, and we find that also B_{n+1} is true. What is the influence of this additional evidence on our confidence in A? Of course,

$$\frac{\begin{array}{c} A \text{ implies } B_{n+1} \\ B_{n+1} \text{ is true} \end{array}}{A \text{ more credible}}$$

Yet how strong is the conclusion? That depends on the credibility of B_{n+1} without A as we have seen in the foregoing section.

Now we may have had a good reason for believing in B_{n+1} before it was verified even under the assumption that A is not true. We have seen previously that $B_1, B_2, \ldots, B_n$ are true. If B_{n+1} is very similar to $B_1, B_2, \ldots, B_n$, we may foresee *by analogy* that also B_{n+1} will be true. If B_{n+1} is very different from $B_1, B_2, \ldots, B_n$, it is not supported by such analogy and we may have very little reason to believe in B_{n+1} without A. Therefore, *the strength of the additional confidence resulting from an additional verification increases when the analogy of the newly verified consequence with the previously verified consequences decreases.*

This expresses essentially the same thing as the complementary patterns formulated in sect. 12.2, but perhaps a little better. In fact, we may regard the explicit mention of analogy as an advantage.

12. On rival conjectures. If there are two different conjectures, A and B, aimed at explaining the same phenomenon, we regard them as opposed to each other even if they are not proved to be logically incompatible. These conjectures A and B may or may not be incompatible, but one of them tends to render the other superfluous. This is enough opposition, and we regard A and B as *rival* conjectures.

There are cases in which we treat rival conjectures almost *as if* they were incompatible. For example, we have two rival conjectures A and B but, in spite of some effort, we cannot think of a third conjecture explaining the same phenomenon; then each of the two conjectures A and B is the "unique obvious rival" of the other. A short schematic illustration may clarify the meaning of the term.

Let us say that A is the emission theory of light that goes back to Newton and that B is the wave theory of light that originated with Huyghens. Let us also imagine that we discuss these matters in the time after Newton and Huyghens, but before Young and Fresnel when, in fact, much inconclusive discussion of these theories took place. Nobody showed, or pretended to show, that these two theories are logically incompatible, and still less that they are the only logically possible alternatives; but there were no other theories of light prominently in view, although the physicists had ample opportunity to invent such theories: each theory was the unique obvious rival of the other. And so any argument that seemed to speak against one of the two rival theories was readily interpreted as speaking for the other.

In general, the relation between rival conjectures is similar to the relation between rivals in any other kind of competition. If you compete for a prize, the weakening of the position of any of your rivals means some strengthening of your position. You do not gain much by a slight setback to one of your many obscure rivals. You gain more if such setback occurs to a dangerous rival. You gain still more if your most dangerous rival drops out of the race. If you have a unique obvious rival, any weakening or strengthening of his position influences your position appreciably. And something similar happens between competing conjectures. There is a pattern of plausible reasoning which we attempt to make somewhat more explicit in Table II.

Table II

A incompatible with B B false	A incompatible with B B less credible
A more credible	A somewhat more credible
A rival of B B false	A rival of B B less credible
A a little more credible	A very little more credible

The disposition of Table II is almost self-explanatory. This Table contains four patterns arranged in two rows and two columns. The first row contains two patterns already considered; see sect. 3, the end of sect. 6, and the last row of Table I. In passing from the first row to the second row, we weaken the first premise; in fact we substitute for a clear relation of formal logic between A and B a somewhat diffuse relation which, however, makes some sense in practice. This weakening of the first premise renders the conclusion correspondingly weaker, as the verbal expression attempts to convey. In passing from the first column to the second column we weaken the second premise, which renders the conclusion correspondingly weaker. The pattern in the southeast corner has no premise that would make sense in demonstrative logic and its conclusion is the weakest.

It is important to emphasize that the verbal expressions used are slightly misleading. In fact, the specifications added to "credible" ("somewhat," "a little," "very little") should not convey any *absolute*, only a *relative*, degree of credibility. They indicate only the change in strength as we pass from one row to the other, or from one column to the other. Even the weakest of the four patterns may yield a weighty conclusion if the conviction that the conjecture A has no other dangerous rival than B is strong enough. In fact, this pattern will play some role in the next chapter.

13. On judicial proof. The reasoning by which a tribunal arrives at its decisions may be compared with the inductive reasoning by which the naturalist supports his generalizations. Such comparisons have been already offered and debated by authorities on legal procedure.[5] Let us begin the discussion of this interesting point by considering an example.

(1) The manager of a popular restaurant that is kept open to late hours returned to his suburban home, as usual, well after midnight. As he left his car to open the door of his garage, he was held up and robbed by two masked individuals. The police, searching the premises, found a dark grey rag in the front yard of the victim; the rag might have been used by one of the holdup men for covering his face. The police questioned several persons in the nearby town. One of the men questioned had an overcoat with a big hole in the lining, but otherwise in good condition. The rag found in the front yard of the victim of the holdup was of the same material as the lining and fitted into the hole exactly. The man with the overcoat was arrested and charged with participation in the holdup.

(2) Many of us may feel that such a charge was amply justified by the related circumstances. But why? What is the underlying idea?

The charge is not a statement of facts, but the expression of a suspicion, of a *conjecture*:

A. The man with the overcoat participated in the holdup.

---

[5] J. H. Wigmore, *The principles of judicial proof*, Boston, 1913; cf. p. 9–12, 15–17.

Such an official charge, however, should not be a gratuitous conjecture, but supported by relevant facts. The conjecture A is supported by the fact B. The rag found in the front yard of the victim of the holdup is of the same material as, and fits precisely into the hole of, the lining of the overcoat of the accused.

Yet why do we regard B as a justification for A? We should not forget that A is just a conjecture: it can be true or false. If we wish to act fairly, we have to consider carefully both possibilities.

If A is true, B is readily understandable. We can easily imagine that a man in urgent need of a mask and not within reach of more suitable materials cuts out a piece from the lining of his overcoat. In a hurry to get away after the criminal act, such a man may lose his mask. Or, in a fright, he may even throw away his mask there and then instead of pocketing it and throwing it away at a safer place. In short, B with A looks readily credible.

If, however, A is not true, B appears inexplicable. If the man did not participate in a holdup or something, why should he spoil his perfectly good overcoat by cutting out a large piece from the lining? And why should that piece of lining arrive, of all places, upon the scene of a robbery by masked bandits? It could arrive there by mere coincidence, of course, but such a coincidence is hard to believe. In short, B without A is hardly credible.

And so we see that the conclusion that led to the charge against the man with the overcoat has the following pattern:

$$\begin{array}{c} \left\{\begin{array}{l} B \text{ with } A \text{ readily credible} \\ B \text{ without } A \text{ hardly credible} \end{array}\right. \\ B \text{ true} \\ \hline A \text{ more credible} \end{array}$$

Yet this pattern of plausible reasoning is obviously related to another pattern of plausible reasoning that we have discussed before (in sect. 10):

$$\begin{array}{c} \left\{\begin{array}{l} A \text{ implies } B \\ B \text{ without } A \text{ hardly credible} \end{array}\right. \\ B \text{ true} \\ \hline A \text{ very much more credible} \end{array}$$

The difference between the two patterns appears at the very beginning. The premise

$$B \text{ with } A \text{ readily credible}$$

is similar to, but weaker than, the premise

$$A \text{ implies } B$$

which, in fact, could also be worded as "B with A certain." Thus the former pattern (just discovered) appears as a "weakened form" of the latter pattern (introduced in sect. 10) and so eventually as a modification of the fundamental inductive pattern (formulated in sect. 12.1).

The case that led us to the formulation of the new pattern was fairly simple. Let us look at a more complex case.[6]

(3) At the time of the murder, Clarence B. Hiller, with his wife and four children, lived in a two-storey house in Chicago. The bedrooms of the family were on the second floor. At the head of the stairs leading to the second floor a gas light was kept burning at night. Shortly after 2 o'clock on Monday morning, Mrs. Hiller was awakened and noticed that this light was out. She awoke her husband and he went in his nightgown to the head of the stairway where he encountered an intruder. The men fought and in the struggle both fell to the foot of the stairway, where Hiller was shot twice; he died in a few moments. The shooting occurred about 2.25 a.m.

Just a little before the shooting one of Hiller's daughters had seen a man at the door of her bedroom, holding a match so that his face could not be seen. She was not frightened because her father used to get up at night and see if the children were all right. No one else in the family saw the intruder.

About three-quarters of a mile from Hiller's house there is a streetcar stop. Early in the morning on which the murder occurred, four police officers, who had gone off duty in the neighborhood shortly before, were sitting on a bench at this stop waiting for the streetcar. About 2.38 a.m. they saw a man approaching from a direction from which the bench could not be easily seen. The officers spoke to the man, but he continued walking with his right hand in his pocket. The officers stopped the man and searched him. There was a loaded revolver in the pocket; he was perspiring; fresh blood appeared at different places of his clothing; there was a slight wound on his left forearm, bleeding slightly. The officers (who did not know at this time of the murder) brought the man to the police station where he was examined. This man—we shall call him the defendant—was later charged with Mr. Hiller's murder.

The court had to examine and, after examination, to deny or uphold the accusation, that is, the following conjecture advanced by the prosecution:

A. The defendant shot and killed Mr. Hiller.

We survey the salient points of the evidence put forward in support of the conjecture *A*.

B_1. There was burned powder in two chambers of the cylinder of the revolver found on the defendant when he was arrested, and there was the smell of fresh smoke. In the judgment of the police officers the revolver had been fired twice within an hour before the arrest. The five cartridges with which the revolver was loaded bore exactly the same factory markings as

---

[6] Concerning the following case the reader should consult the findings of the Supreme Court of Illinois, almost fully reprinted in Wigmore, l.c. footnote 5, p. 83–88.

three undischarged cartridges found in the hallway of the Hiller house near the dead body.

B_2. The intruder entered the Hiller house through a rear window of the kitchen from which he first removed the screen. A person entering this window could support himself on the railing of the porch. On this railing which was freshly painted there was the imprint of four fingers of a left hand. Two employees of the identification bureau of the Chicago police testified that, in their judgment, the imprints on the railing were identical with the defendant's fingerprints.

B_3. Two experts not belonging to the Chicago police expressed the same opinion concerning the fingerprints. (One was an inspector of the Dominion Police at Ottawa, Canada; the other a former expert of the federal government in Washington, D.C.)

B_4. About 2.00 a.m., just before the shooting of Mr. Hiller, a prowler entered a house separated by a vacant lot from the Hiller house. Two women saw a man in the door of their bedroom with a lighted match over his head. Both women testified that this prowler was of the same size and build as the defendant. One of the women remembered also that the prowler wore a light-colored shirt and figured suspenders. Having inspected the shirt and the suspenders of the defendant exhibited in court, the witness testified that in her opinion the defendant was the same man that she saw on that night in the door.

B_5. The defendant gave a false name and a false address when he was arrested and he denied that he had ever been arrested before. In fact, he had been sentenced before on a charge of burglary, paroled, returned to the penitentiary for a violation of the parole, and released on parole a second time about six weeks before the night of the murder. He bought the revolver under a false name about two weeks after his second parole, pawned it, got it back, pawned it again and got it back a second time five hours before the shooting.

B_6. The defendant was unable to explain consistently the blood on his clothing, or the wound on his left forearm, or his whereabouts on the night of the shooting. Concerning his whereabouts he told two different stories, one after his arrest and another in court. The people whom he first asserted having visited that night denied that he called on them. The defendant then told the court that he visited a saloon, but no witness was found to corroborate this.

(4) All the facts, events, and circumstances related under the headings $B_1, B_2, \ldots B_6$ are readily understandable *if* the accusation A is true. They all support A, but the weight of such support is not the same in all cases. Some of these facts would be explicable even if A was not true. Some others, however, would appear as miraculous coincidences if A was not true.

That the cartridges found in the revolver of the defendant are of the same make as the cartridges found next to the body of the victim proves little in itself, if this make is a usual make sold by all gunsmiths. Yet it proves a lot that just as many cartridges have been fired from this revolver as shots have been fired at the victim and within the same hour; it is difficult to explain away such a coincidence. The agreement of the fingerprints on the railing with the fingerprints of the defendant would be considered in itself as an almost decisive proof nowadays, but was not yet considered so at the time of the trial, in 1911. That the defendant lied about his name, address, and criminal past when he was arrested does not prove much; such a lie is understandable if the accusation A is true, but also understandable if it is not true: the man would prefer to be let alone by the police anyway. Yet it weighs heavily that the defendant was not able to explain consistently his whereabouts on that fatal night. He must have known that the point is important, and his counsel certainly knew about the importance of an alibi. If the accusation A was untrue and the defendant passed the night harmlessly or committing some minor crime, why did he not say so right away, or at least before it was too late?

All the details mentioned are readily understandable if the accusation A is true. Yet the coincidence of so many details appears as inexplicable if the accusation A is not true; it is extremely hard to believe that so many coincidences are due to mere chance. At any rate, the defense failed to propose a consistent explanation of the evidence submitted.

There was, of course, more evidence before the jury than related here, and there was one that no description can adequately render: the behavior of the defendant and the witnesses. The jury convicted the defendant of murder and the State Supreme Court upheld the conviction. We quote the last sentence of the opinion of the Chief Justice: "No one of these circumstances, considered alone, would be conclusive of his [the defendant's] guilt, but when all the facts and circumstances introduced in evidence are considered together, the jury were justified in believing that a verdict of guilty should follow as a logical sequence."[7]

(5) The statements $B_1, B_2, \ldots B_6$ listed under (3) fit pretty well the pattern introduced under (2). They fit even better another pattern which differs from it in just one point:

$$\frac{\begin{cases} B \text{ with } A \text{ readily credible} \\ B \text{ without } A \text{ less readily credible} \end{cases} \quad \\ B \text{ true}}{A \text{ more credible}}$$

Each of the statements $B_1, B_2, \ldots B_6$ can be meaningfully substituted for B in this pattern in which, of course, A has to be interpreted as the accusation. The statements $B_1, B_2, \ldots B_6$ are compound statements; they have

[7] Cf. Wigmore, *loc. cit.* 5, p. 88.

portions (some of which we have emphasized under (4)) each of which can be regarded in itself as relevant evidence: each such portion can also be meaningfully substituted for B in the above pattern. If we recall our discussion under (4) we may realize that, of course, the less easily credible B is without A, the stronger is the conclusion.

If we visualize how the successive witnesses unfolded the mass of evidence before the jury in the course of the proceedings, we may see more clearly the analogous roles of plausible reasoning in such proceedings and in a scientific inquiry in which several consequences of a conjecture are successively tested. (Cf. especially sect. 12.2.)

(6) The foregoing consideration clearly suggests a compound pattern of plausible reasoning that is related to the pattern stated under (5) just as the compound pattern introduced in sect. 12.2 is related to the fundamental inductive pattern of sect. 12.1. I do not enter upon this matter here; a reader more versed in the doctrine of legal evidence could take it up with more impressive examples and interpretations, but I will add one more illustration of the pattern.[8]

As Columbus and his companions sailed westward across an unknown ocean they were cheered whenever they saw birds. They regarded birds as a favorable sign, indicating the nearness of land. Although in this they were repeatedly disappointed, the underlying reasoning seems to me quite correct. Stated at length, this reasoning runs as follows:

{ When the ship is near the land, we often see birds.
{ When the ship is far from the land, we see birds less often.

Now we see birds.

Therefore, it becomes more credible that we are near the land.

This reasoning fits exactly the pattern formulated under (5): the presence of birds is regarded as circumstantial evidence for the nearness of land. Columbus' crew saw several birds on Thursday, the 11th of October, 1492, and the next day they discovered the first island of a New World.

The reader may notice that the pattern illustrated underlies much of our everyday reasoning.

EXAMPLES AND COMMENTS ON CHAPTER XIII

First Part

1. Following the method of sect. 4 (6), derive the demonstrative pattern mentioned in sect. 3 from the demonstrative pattern mentioned in sect. 1

[8] *How to Solve It*, pp. 212–221.

2. Supply the details of the demonstration sketched in sect. 5: derive the heuristic pattern of sect. 2 from the heuristic pattern of sect. 1.

3. Derive the heuristic pattern of sect. 3 from the heuristic pattern of sect. 1.

4. In a crossword-puzzle we have to find a word with 9 letters, and the clue is: "Disagreeable form of tiredness."[9]

The condition that the unknown word has to satisfy is ambiguously stated, of course. After a few unsuccessful attempts we may observe that "tiredness" has 9 letters, just as many as the unknown word, and this may lead us to the following conjecture:

A. The unknown word means "disagreeable" and is an *anagram* of TIREDNESS.

(Anagram means a rearrangement of the letters of the given word into a new word.) This conjecture *A* may appear quite likely. (In fact, "form of" may suggest "anagram of" in crossword jargon.) Tackling other unknown words of the puzzle, we find quite plausible solutions for two of them crossing the nine-letter word mentioned, for which we obtain two possible letters, placed as the following diagram indicates:

— — — — — — T — R.

We may regard this as evidence for our conjecture *A*.

(a) Why? Point out the appropriate pattern.

(b) Try to find the nine-letter word required. [In doing so, you have a natural opportunity to weigh the evidence for *A*. Here are a few helpful questions: Which letter is the most likely between T and R? How could the vowels E E I be placed? See also *How to Solve It*, pp. 147–149.]

5. Let us take up once more the court case already considered in sect. 12.3 and (more fully) in sect. 10. Let us consider again the charge (the *Factum Probandum*, the fact to prove, of the prosecution):

A. The defendant blew up the yacht.

Yet let us change our notation in another point: we consider here the statement

B. The defendant bought dynamite in such and such an amount, in such and such a shop, on such and such a day.

The change is that *B* denotes here not a general statement, but a specific fact. (Courts prefer, or should prefer, to deal with facts as distinctly specified as possible.) We again take *B* as proved. (Thus, *B* is a *Factum Probans*, a fact supporting the proof.)

Such change of notation as we have introduced cannot change the force of the argument. Yet what is now the pattern?

6. The defendants are a contractor and a public official. One was charged with giving, the other with accepting, a bribe. There was a specific charge:

---

[9] *The Manchester Guardian Weekly*, November 29, 1951.

the accusation asserted that the down payment for the new car of the official came from the contractor's pocket. One of the witnesses for the prosecution was a motor car dealer; he testified that he received $875 on November 29 as down payment on the official's car. Another witness was the manager of a local bank; he testified that $875 was withdrawn on November 27 (of the same year) from a usually inactive joint account of the contractor and his wife; the receipt was signed by the wife. These facts were not challenged by the defense.

What would you regard as a strong point in this evidence? Name the appropriate pattern.

7. The Blacks, the Whites, and the Greens live on the same street in Suburbia. The Blacks and the Whites are neighbors and the Greens live just opposite. One evening Mr. Black and Mrs. White had a long conversation over the fence. It was rather dark, yet Mrs. Green did not fail to observe the conversation and jumped to the conclusion—you know which conclusion: that favorite conjecture of Mrs. Green.

Unfortunately, there is not much chance to stem the flood of gossip started by Mrs. Green. If, however, taking a terrific risk, you wished to constitute yourself counsel for the defense of innocent people maligned by Mrs. Green, I can give you a fact: the Whites, who wished to move nearer to Mr. White's office for a long time, signed a lease for a house that belongs to Mr. Black's uncle, and this happened a few days after that conversation. Use this fact.

What is your defense and what is the pattern?

8. The prosecution tries to prove:

A. The defendant knew, and was capable of recognizing, the victim at the time of the crime.

The prosecution supports this by the undenied fact:

C. The defendant and the victim were both employed by the same firm for several months three years before the crime was committed.

Thus *A* is the Factum Probandum and *C* is advanced as Factum Probans. What is the pattern? [The notation is devised to help you. Is the size of the firm relevant?]

9. *On inductive research in mathematics and in the physical sciences.* A difference between "mathematical situations" and "physical situations" that seems to be important from the standpoint of plausible reasoning has been pointed out in sect. 6. There seem to be other differences of this kind, and one should be discussed here.

Coulomb discovered that the force between electrified bodies varies inversely as the square of the distance. He supported this law of the inverse square by direct experiments with the torsionbalance. Coulomb's experiments were delicate and the discrepancy between his theoretical and experimental numbers is considerable. We cannot help suspecting that without the powerful analogy of Newton's law (the law of the inverse square in gravitational attractions), neither Coulomb himself nor his contemporaries would have considered his experiments with the torsion-balance as conclusive.

Cavendish discovered the law of the inverse square in electrical attractions and repulsions independently of Coulomb. (Cavendish's researches were not published in his lifetime, and Coulomb's priority is incontestable.) Yet Cavendish devised a more subtle experiment to support the law. We need not discuss the details of his method,[10] only one feature of which is essential here: Cavendish considers the possibility that the intensity of the force is not r^{-2} (r is the distance of the electric charges) but, more generally, $r^{-\alpha}$ where α is some positive constant. His experiment shows that $\alpha - 2$ cannot exceed in absolute value a certain numerical fraction.

Coulomb's experimental investigation is pretty similar to an inductive investigation in mathematics: he confronts particular consequences of a conjectural physical law with the observations, as a mathematician would confront particular consequences of a conjectural number-theoretic law with the observations. Analogy may play an important rôle in the choice of the conjectural law, here and there. Yet Cavendish's experimental investigation is of a different character; he does not consider only one conjectural law (the law r^{-2}) but several conjectural laws (the laws $r^{-\alpha}$). These laws are different (different laws of electrical attraction correspond to different values of the parameter α) but they are related, they belong to the same "family" of laws. Cavendish confronts a whole family of laws with the observations and tries to pick out the law that agrees best with them.

This is the most characteristic difference between the two investigations: one aims at *one* conjecture, the other at a *family* of conjectures. The first compares the observations with the consequences of one conjecture, the other compares them with the consequences of several conjectures simultaneously. The first tries to judge on the basis of such comparison whether the proposed conjecture is acceptable or not, the other tries to find out the most acceptable (or least unacceptable) conjecture. The first kind of inductive investigation is widely practised in mathematics and is not unusual in the physical sciences. The second kind of inductive investigation is widely practised in the physical sciences, but we very seldom meet with it in mathematics.

In fact, the most typical kind of physical experiment aims at measuring some physical constant, at determining its value, as Cavendish's experiment aims at determining the value of the exponent α. In mathematics we could regard this or that investigation as an inductive research aiming at the

[10] Cf. J. C. Maxwell, *A treatise on electricity and magnetism*, 2nd ed. (1881), vol. 1, p. 76–82.

determination of some mathematical constant, but such investigations are quite exceptional.[11]

10. *Tentative general formulations.* Newton's often repeated "Hypotheses non fingo" is unilateral. It would be a mistake to interpret it as "Beware of conjectures": such advice, if followed, would ruin inductive investigation. Better advice is: Be quick in shaping conjectures; go slow in accepting them. Still better is Faraday's word: "The philosopher should be a man willing to listen to every suggestion, but determined to judge for himself." Of course, the philosopher that Faraday had in mind cultivates the experimental, not the traditional, philosophy.

The present study intends to be an "inductive" investigation of plausible reasoning. I present here, without inhibition, a few tentative generalizations. They apply to a few forms of plausible reasoning, but even their formulation should be carefully scrutinized before any attempt at a further extension.

(1) *Monotonicity.* The considerations of sect. 6 may suggest a rule: "The conclusion of a plausible inference varies monotonically, when one of its premises varies monotonically." This fits the case considered in sect. 6 and a few more cases some of which will be presently considered.

(2) *Continuity.* We shall need a term of demonstrative logic. We say that

$$A \text{ and } B \text{ are equivalent}$$

when A and B imply each other mutually, that is, when A follows from B and also B follows from A. If A and B are equivalent, we may not know at present whether A or B is true, but we know that only two cases are possible: either both are true or both are false; A and B stand or fall together. A descriptive symbolic expression for the equivalence of A and B is:

$$A \rightleftarrows B.$$

The two arrows suggest that we can pass from the truth of any one of the two statements A and B to the truth of the other.

In sect. 10 we considered a suggestive connection between two statements A and B. We considered the logical relation

$$A \text{ implies } B$$

along with the credibility of B without A. Let us imagine that this credibility varies monotonicallv· B becomes less and less credible without A. In the limit when B becomes impossible without A, the truth of B implies that of A. Yet we supposed that the truth of A implies that of B and so, in the limit, A and B imply each other mutually, become equivalent.

---

[11] By the way, Cavendish's experiment has an even broader scope: it tends to show that the law r^{-2} is more acceptable than any other law $\varphi(r)$, without restricting the function $\varphi(r)$ to the form $r^{-\alpha}$; cf. Maxwell, *ibid.*, p. 76–82.

Let us observe now how the variation described affects our fundamental inductive pattern:

$$\frac{A \text{ implies } B \qquad B \text{ true}}{A \text{ more credible}}$$

We assume that the second premise remains unchanged when the first premise varies as described. As B becomes less and less credible without A, the conjecture A becomes more and more credible by the verification of its consequence B. That is, the conclusion becomes stronger, gains weight. In the limit the conclusion becomes "A is true" and so our pattern of plausible inference becomes in the limit the following (obvious) pattern of demonstrative inference

$$\frac{A \text{ and } B \text{ equivalent} \qquad B \text{ true}}{A \text{ true}}$$

In short, our pattern of plausible inference has a "limiting form," which is a pattern of demonstrative inference. As the premises of the plausible inference "tend" to the corresponding premises of the limiting form, the plausible conclusion "approaches" its extreme limiting strength. Still shorter: there is a continuous transition from the heuristic pattern to a demonstrative pattern.

Most of this description fits a few more cases. Some of these are displayed in Table III. The symbols and abbreviations of Table I, explained in sect. 7, are used in Table III. Also the symbol $\leftrightarrows$, explained above, is used. Instead of non-B, defined in sect. 4, the shorter symbol $\bar{B}$ is used.

Table III

Approaching:	$A \to B$ B true ——— A more cr.	$A \to B$ B more cr. ——— A s. more cr.	$A \mid B$ B false ——— A more cr.	$A \mid B$ B less cr. ——— A s. more cr.
Limiting:	$A \rightleftarrows B$ B true ——— A true	$A \leftrightarrows B$ B more cr. ——— A more cr.	$A \rightleftarrows \bar{B}$ B false ——— A true	$A \rightleftarrows \bar{B}$ B less cr. ——— A more cr.

(3) *Plausible from demonstrative?* Table III may yield another suggestion. The limiting patterns in this Table are much more obvious than the approaching patterns. Two of the limiting patterns are demonstrative, and the two others, although not purely demonstrative, are hardly questionable. The more debatable approaching patterns which are all patterns of plausible

inference, seem to arise from the corresponding limiting patterns by a uniform "weakening" process: stronger statements as

$$A \rightleftarrows B, \quad A \text{ true}, \quad A \text{ more cr.}$$

are systematically replaced by corresponding weaker statements as

$$A \rightarrow B, \quad A \text{ more cr.}, \quad A \text{ s. more cr.}$$

Could *all* forms of plausible inference be linked in some analogous way to forms of demonstrative, or almost demonstrative, inference?

Second Part

Ex. 11 should be read first: it introduces (and excuses) what follows.

11. *More personal, more complex.* In the foregoing I did not discuss conjectures from my own published mathematical work. This is an omission since, after all, I cannot know any mathematician more intimately than myself. This omission could even appear suspicious to some readers. I do not think that such suspicion is justified. The reason for not discussing more complex specialized topics from my own research is not lack of frankness, but just the complexity and specialization of the topics; I thought it better to discuss simpler topics of more general interest.

The following ex. 12–19 suppose considerably more advanced knowledge than the bulk of the book. They are taken from my own research. I try to offer a representative sample. I include some conjectures that have appeared in print before, and others that have not. I include some conjectures from my "naïve" early work, when I had not begun to think explicitly about the subject of plausible reasoning, and I include conjectures from my later, less naïve work. Ex. 12 dates from my naïve days; it tells of the heuristic grounds that led me to a result. Ex. 13, 14, 15, and 16 deal with formerly published conjectures from my naïve years, ex. 17 with a formerly published conjecture from my less naïve years, and ex. 18 and 19 with conjectures that have not been printed before.[12]

---

[12] Papers by the author of this book are quoted without name in this footnote; "cf." introduces the page on which the conjecture is stated (sometimes in form of a question); a paper quoted with name brought the first proof of the conjecture discussed. *Ex.* 12: *Rendiconti, Circolo Matematico di Palermo*, vol. 34 (1912), p. 89–120. *Ex.* 13: *L'Intermédiaire des Mathématiciens*, vol. 21 (1914), p. 27, qu. 4340; G. Szegö, *Math. Annalen*, vol. 76 (1915), p. 490–503. *Ex.* 14: *Math. Annalen*, vol. 77 (1916), p. 497–513; cf. p. 510; F. Carlson, *Math. Zeitschrift*, vol. 9 (1921), p. 1–13. *Ex.* 15: *L'Intermédiaire des Mathématiciens*, vol. 20 (1913), p. 145–146, qu. 4240; G. Szegö, *Math. Zeitschrift*, vol. 13 (1922), p. 38. See also *Journal für die reine und angewandte Math.*, vol. 158 (1927), p. 6–18. *Ex.* 16: *Jahresbericht der Deutschen Math. Vereinigung*, vol. 28 (1919), p. 31–40; cf. p. 38. *Ex.* 17: *Proceedings of the National Academy of Sciences*, vol. 33 (1947), p. 218–221; cf. p. 219. *Ex.* 19: *Journal für die reine und angewandte Math.*, vol. 151 (1921), p. 1–31; see theorem I, p. 3.

I should add that even in my naïve years I was impressed and somewhat puzzled by the strength of the confidence with which some of my own conjectures inspired me, and I wondered what sort of reasons might underlie such confidence. The following sentences express fairly well my early views about the source of new conjectures.

"There is a straight line that joins two given points. A new theorem is often a generalization that joins two extreme particular cases, and is obtained by a sort of linear interpolation. There is a straight line with a given direction through a given point. A new theorem is often conceived in the fortunate moment when a general direction of inquiry meets with an appropriate particular case. A new theorem may also result from drawing a parallel."[13]

12. *There is a straight line that joins two given points.* The numbers $A_n^{(k)}$ disposed in the infinite square

$$\begin{array}{cccccccc} A_0^{(1)} & A_1^{(1)} & A_2^{(1)} & A_3^{(1)} & \ldots & A_n^{(1)} & \ldots \\ A_0^{(2)} & A_1^{(2)} & A_2^{(2)} & A_3^{(2)} & \ldots & A_n^{(2)} & \ldots \\ \cdot & \cdot & \cdot & \cdot & \cdot & \cdot & \cdot \\ A_0^{(k)} & A_1^{(k)} & A_2^{(k)} & A_3^{(k)} & \ldots & A_n^{(k)} & \ldots \\ \cdot & \cdot & \cdot & \cdot & \cdot & \cdot & \cdot \end{array}$$

are connected with a function $f(x)$. (They are derived from the expansion of $f(x)$ in powers of x; both $A_n^{(k)}$ and $f(x)$ are real.) E. Laguerre discovered that $V(k)$, the number of changes of sign in the kth row $A_0^{(k)}, A_1^{(k)}, A_2^{(k)}, \ldots$ of the array, is connected with R, the number of roots of the equation $f(x) = 0$ between 0 and 1 (extremities excluded):

$$V(k) \geqq R,$$

and $V(k)$ can only decrease or remain unchanged as k increases. Will the never increasing $V(k)$ ultimately attain R? Laguerre proposed this question, but left it unsolved. M. Fekete proved that $V(k)$ finally attains R provided that $R = 0$, and this particular case suggested the conjecture that $V(k)$ always attains R. I was lucky enough to observe another connection between the numbers $A_n^{(k)}$ and the function $f(x)$: there are certain positive numbers $B_n^{(k)}$ (independent of the function $f(x)$) such that

$$\lim_{k\to\infty} A_n^{(k)}/B_n^{(k)} = f(0) \text{ for fixed } n,$$

$$\lim_{n\to\infty} A_n^{(k)}/B_n^{(k)} = f(1) \text{ for fixed } k.$$

That is, the vertical direction in the infinite square array is connected with $f(0)$, the horizontal direction with $f(1)$. These two extreme directions,

---

[13] Cf. G. Polya and G. Szegö, *Analysis*, vol. 1, p. VI.

the vertical and the horizontal, may remind you of *two extreme points that demand to be joined*: what about the intermediate oblique directions? Just for a moment, take the conjecture that we aim at proving for granted: it implies some connection between the intermediate directions and the values that $f(x)$ takes when x varies between 0 and 1. This suggests that $A_n^{(k)}/B_n^{(k)}$ may tend to $f(x)$ where x is somehow connected with the limit of n/k. In fact, I found finally, and it was even easy to prove, that

$$\lim_{k,n\to\infty} A_n^{(k)}/B_n^{(k)} = f(x) \text{ if } \lim n/(n+k) = x.$$

This relation turned out the key to the solution of Laguerre's problem.

13. *There is a straight line with a given direction through a given point. Drawing a parallel.* In conjunction with the Fourier series of a positive function $f(x)$,

$$f(x) = a_0 + 2\sum_1^\infty (a_n \cos nx + b_n \sin nx),$$

O. Toeplitz considered the equation in λ

$$(*)\qquad \begin{vmatrix} a_0 - \lambda & a_1 - ib_1 & \dots & a_{n-1} - ib_{n-1} \\ a_1 + ib_1 & a_0 - \lambda & \dots & a_{n-2} - ib_{n-2} \\ & & \cdot & \\ a_{n-1} + ib_{n-1} & a_{n-2} + ib_{n-2} & \dots & a_0 - \lambda \end{vmatrix} = 0.$$

His research revealed that the n roots of this equation

$$\lambda_{n1},\quad \lambda_{n2},\quad \lambda_{n3},\quad \dots\quad \lambda_{nn}$$

"imitate" the n equidistant values of the function $f(x)$

$$f\left(\frac{2\pi}{n}\right),\ f\left(\frac{4\pi}{n}\right),\ f\left(\frac{6\pi}{n}\right), \dots f\left(\frac{2n\pi}{n}\right).$$

For example, the arithmetic mean of the n roots

$$\frac{\lambda_{n1} + \lambda_{n2} + \dots + \lambda_{nn}}{n} = a_0$$

corresponds to

$$\lim_{n\to\infty}\left[f\left(\frac{2\pi}{n}\right) + f\left(\frac{4\pi}{n}\right) + \dots + f\left(\frac{2n\pi}{n}\right)\right]\frac{1}{n} = \frac{1}{2\pi}\int_0^{2\pi} f(x)\,dx = a_0.$$

Let us try to *draw a parallel*: the geometric mean of the n roots

$$[\lambda_{n1}\lambda_{n2}\dots\lambda_{nn}]^{1/n} = D_n^{1/n}$$

where D_n denotes the determinant with n rows that we obtain in setting $\lambda = 0$ on the left-hand side of the equation (*). This may correspond to

$$(**)\qquad \lim_{n\to\infty}\left[f\left(\frac{2\pi}{n}\right)f\left(\frac{4\pi}{n}\right)\dots f\left(\frac{2n\pi}{n}\right)\right]^{1/n} = e^{\frac{1}{2\pi}\int_0^{2\pi}\log f(x)\,dx}$$

At this stage it is natural to look for a convenient particular case. For the particular function

$$f(x) = a_0 + 2a_1 \cos x + 2b_1 \sin x$$

there is no difficulty in computing D_n and $\lim_{n\to\infty} D_n^{1/n}$, and this limit turns out to be equal to the value (**): *a general direction of inquiry met with an appropriate particular case,* and it would have been difficult not to state the conjecture: for any positive function $f(x)$

$$\lim_{n\to\infty} D_n^{1/n} = e^{\frac{1}{2\pi}\int_0^{2\pi} \log f(x)\,dx}$$

14. *The most obvious case may be the only possible case.* If the coefficients $a_0, a_1, a_2, \ldots a_n, \ldots$ of the power series

$$a_0 + a_1 z + a_2 z^2 + \ldots + a_n z^n + \ldots$$

are integers, and an infinity of these integers are different from 0, the series obviously diverges at the point $z = 1$ since its general term does not tend to 0. Therefore the radius of convergence of such a power series is $\leqq 1$. The extreme value 1 of the radius of convergence may be attained. An obvious example is the series

$$1 + z + z^2 + \ldots + z^n + \ldots ;$$

it represents a function of extremely simple analytic character, the rational function $1/(1 - z)$. Another example is

$$z + z^2 + z^6 + \ldots + z^{n!} + \ldots ;$$

it represents a function of extremely complex analytic character, a non-continuable function. (This is the most familiar example of a series for which the circle of convergence is a singular line.) These two obvious examples are of opposite nature. Yet, surprisingly, any power series with integral coefficients and radius of convergence 1, the analytic nature of which could be ascertained, turned out to be similar to one or the other of these opposite examples: it represented either a rational function or a non-continuable function. Theorems by E. Borel and P. Fatou showed that extensive classes of functions of intermediate nature cannot be represented by such a series. As the author succeeded in proving similar theorems and in handling more examples, he unearthed mounting evidence for a conjecture: "The *most obvious* continuable analytic functions represented by a power series with integral coefficients and the radius of convergence 1, certain rational functions, are the *only such* functions." In other words: "If the radius of convergence of a power series with integral coefficients attains the extreme value 1, the function represented is necessarily of an extreme character: it is either extremely simple, a rational function, or extremely complicated, a non-continuable function."

15. *Setting the fashion. The power of words.* This example has various aspects which I shall try to discuss one after the other.

(1) E. Laguerre discovered several sequences of real numbers

$$\alpha_0, \alpha_1, \alpha_2, \ldots \alpha_n$$

with the following curious property: if the (otherwise arbitrary) equation of degree n

$$a_0 + a_1 x + a_2 x^2 + \ldots + a_n x^n = 0 \tag{I}$$

has real roots only, also the equation

$$a_0\alpha_0 + a_1\alpha_1 x + a_2\alpha_2 x^2 + \ldots + a_n\alpha_n x^n = 0 \tag{II}$$

(into which the sequence transforms (I)) will have real roots only. Laguerre proposed, but left unsolved, the problem: to find a simple necessary and sufficient condition that characterizes this kind of sequences.

It is easy to find necessary conditions: we just apply a sequence of the desired kind to any equation that is known to have real roots only, and then we obtain a transformed equation all roots of which are necessarily real. For example, applying the sequence $\alpha_0, \alpha_1, \ldots \alpha_n$ to the equations

$$1 - x^2 = 0, \qquad x^2 - x^4 = 0, \qquad x^4 - x^6 = 0, \qquad \ldots$$

which, obviously, have real roots only, we easily find that $\alpha_0, \alpha_2, \alpha_4, \alpha_6, \ldots$ are necessarily of the *same sign*, all positive, or all negative (in the broad sense: 0 is not excluded). Applying the same sequence to the equations

$$x - x^3 = 0, \qquad x^3 - x^5 = 0, \qquad x^5 - x^7 = 0, \qquad \ldots$$

we find that also $\alpha_1, \alpha_3, \alpha_5, \alpha_7, \ldots$ are necessarily of the same sign. Bearing these remarks in mind and applying the sequence to the equation

$$1 + \binom{n}{1} x + \binom{n}{2} x^2 + \ldots + x^n = 0 \tag{III}$$

(all n roots of which are equal to -1) we obtain the necessary condition that the roots of the equation

$$\alpha_0 + \binom{n}{1} \alpha_1 x + \binom{n}{2} \alpha_2 x^2 + \ldots + \alpha_n x^n = 0 \tag{IV}$$

are all real and of the same sign.

We obtained this last condition by accumulating several necessary conditions. Their accumulation should be rather strong: *is it strong enough to form a sufficient condition?* If it were so, the following curious proposition would hold: "If both equations (I) and (IV) have real roots only, and the roots of (IV) are all of the same sign, also the equation (II) has only real roots."

(2) This conjecture came rather early to my mind, but I could not trust it: it looked too strange. In the case $n = 2$, however, the conjecture was easily verified (the case $n = 1$ is completely trivial).

Yet, by chance, I came across a theorem proved by E. Malo: If the equation (I) has only real roots and the equation

$$\alpha_0 + \alpha_1 x + \alpha_2 x^2 + \ldots + \alpha_n x^n = 0$$

has only real roots of the same sign, the equation (II) must have only real roots. Malo's theorem was clearly analogous to that strange conjecture and made it look much less strange. Moreover, Malo's theorem was a consequence of that conjecture, as I could easily see, and so another broad and important consequence of the conjecture had been verified: the conjecture appeared much stronger.

(3) As it is easily seen, the conjecture can be restated as follows: "If the sequence of positive numbers $\alpha_0, \alpha_1, \ldots \alpha_n$ transforms the equation (III) into an equation with real roots only, it will transform an arbitrary equation with real roots only into an equation of the same nature." In other words, the equation (III) *sets the fashion*: its response to the sequence $\alpha_0, \alpha_1, \ldots \alpha_n$ is imitated by all equations whose roots are all real.

(4) Why does equation (III) set the fashion? Because all its roots coincide. This answer is the "right" answer in some sense. At any rate, as I found out later, equations (or functions) all roots (or zeros) of which coincide play an analogous role in several analogous problems: they set the fashion (they are "*tonangebend*").

Yet, as the conjecture in question was still a conjecture, I resorted to the following "explanation": All roots of the equation (III) are equal to -1. These roots are all crowded into one point of the real axis, they are as close together as possible. In such a situation they are, understandably, the most inclined to jump out of the real axis. Therefore, if the sequence $\alpha_0, \alpha_1, \ldots \alpha_n$ applied to the polynomial $(1 + x)^n$ with the most crowded roots does not succeed in driving out these roots of the real axis, it has still less chance to drive out the less crowded roots of other polynomials.

The logical value of this "explanation" is obviously nil, but that does not imply that its psychological value is also nil. I am convinced that this playful analogy was extremely important for me personally: it helped to keep the conjecture alive for years.

I should mention here that similar quaint verbal formulations have often been connected with my mathematical work. The sentence at the end of ex. 14 is a characteristic instance. Here are two more examples.

For more than two decades I was much interested in Fabry's well known gap-theorem on power series. There were two periods: a first "contemplative" period, and a second "active" period. In the active period I did some work connected with the theorem, and found various proofs, extensions, and analogues to it. In the contemplative period I did practically no work connected with the theorem, I just admired it, and recalled it from time to time in some curious, far fetched formulation as the following: "If it is infinitely improbable that in a power series a coefficient chosen at random be different from 0, then it is not only infinitely probable, but certain, that the power series is not continuable." Obviously, this sentence has neither logical, nor literary merit, but it served me well in keeping my interest alive.

The idea of a certain proof occurred to me fairly clearly, but for several days after that I did not attempt to work out the final shape of the proof. During these days I was obsessed by the word "transplantation." In fact, this word describes the decisive idea of the proof as precisely as it is possible for any single word to describe such a complex thing.

I made up, of course, various explanations for this "power of words" but, perhaps, it is better to wait with explanations till there are more examples.[14]

16. *This is too improbable to be a mere coincidence.* Let f denote the number of prime factors of the integer n, and let us call n "evenly factorized" or "oddly factorized" according as f is even or odd. For example:

$$30 = 2 \times 3 \times 5 \quad \text{is oddly factorized}$$

$$60 = 2 \times 2 \times 3 \times 5 \quad \text{is evenly factorized.}$$

Prime numbers as 2, 3, 5, 7, 11, 13, 17, . . . are oddly factorized, squares as 4, 9, 16, . . . are evenly factorized, and the number 1 has to be regarded as evenly factorized since it has no prime factors and 0 is an even number. Among the twelve first numbers

1	2	3	4	5	6	7	8	9	10	11	12
e	o	o	e	o	e	o	o	e	e	o	o

five are evenly factorized, and seven oddly.

If we look at the succession of e's and o's in the foregoing scheme, we can scarcely detect a simple rule. The two kinds of numbers appear to alternate irregularly, unpredictably, *at random.* The idea of chance occurs to us almost unavoidably: it is natural to think that we would obtain a similar sequence if, instead of taking the trouble to factorize a proposed integer, we just flipped a coin and wrote "e" or "o" according as head or tail turns up.

[14] Cf. J. Hadamard, The psychology of invention in the mathematical field, p. 84–85.

It is also natural to suspect that the coin is "fair," that heads and tails turn up about equally often, that the two kinds of integers, the evenly and the oddly factorized, are equally frequent.

Now it can be proved (the proof is difficult) that among the first n integers about as many are evenly factorized as oddly factorized if n is large. (The ratio tends to 1 as n tends to ∞.) This seems to corroborate our suspicions. Now you would more confidently expect that the evenly and oddly factorized numbers follow each other as a random succession of heads and tails. Expecting so, I started listing for each n which kind of integers, the oddly or the evenly factorized, are in majority among the first n. The first listings looked lopsided. I was surprised and proceeded to larger values of n: it was still lopsided. I was amazed and proceeded to still larger values of n, but it was still lopsided. I got tired of numerical computation as I reached $n = 1500$ and had to admit that there is some observational evidence for the conjecture: "For $n \geqq 2$, the evenly factorized integers are never in majority among the first n integers."

You toss a coin 1500 times. You count how many heads and how many tails you obtain in the first n trials. It can easily happen that you obtain less heads than tails. (Take "less" in the wide sense, as "not more.") It could even happen that you obtain less heads than tails all the time from the second step on, for $n = 2, 3, 4, \ldots 1500$, but this cannot so easily happen. *This is too improbable to be a mere coincidence*, we are tempted to say. Yet the improbable did happen and has been observed as we played heads and tails with the factorized integers. *There must be some reason.* I did what I had to do: I took the conjecture for granted (tentatively, of course) and tried to derive consequences from it. I was lucky enough to observe two points. First, if the new conjecture were true, the truth of a far more important conjecture due to Riemann (on the ζ-function) would necessarily follow. Second, if a little more than the new conjecture were true (if the evenly factorized numbers were definitely in the minority from a certain n upward) another important conjecture due to Gauss (on the classnumber of quadratic forms) would necessarily follow. Both points seemed to speak in favor of the new conjecture.

The conjecture is no longer new, but its fate is still undecided. A. E. Ingham derived consequences from it that may tend to render it less credible. On the other hand, D. H. Lehmer verified it by numerical computation up to $n = 600{,}000$.

17. *Perfecting the analogy.* "Of all solids with a given volume, the sphere has the minimum surface." This is the classical isoperimetric theorem in space, a remarkable physical analogue to which was discovered by H. Poincaré and strictly proved by G. Szegö: "Of all solids with a given volume the sphere has the minimum electrostatic capacity." It is natural to think that there should be further analogous theorems, and I was looking for such a theorem. The field of force around a solid charged with electricity is similar to the field of flow around a solid moving with uniform velocity across an incompressible ideal fluid. (Both fields are sourceless and irrotational, and therefore satisfy the same partial differential equation.) The capacity of the solid in the electrostatic field corresponds roughly to the "virtual mass" of the solid in the hydrodynamic field. (The moving body stirs up the fluid and adds to its own kinetic energy that of the moving fluid; the "virtual mass" is a factor of this additional kinetic energy.) Both the capacity and the virtual mass are connected with the energy of the corresponding field. Yet there is a conspicuous difference: the capacity depends only on the shape and size of the solid, but the virtual mass depends also on the direction of the motion of the solid. In order to render the analogy more perfect, I had to construct a new concept: averaging the virtual mass over all the possible directions, we obtain the *average virtual mass*. And so the conjecture arose: "Of all solids with a given volume, the sphere has the minimum average virtual mass."

The ellipsoid is the only shape for which the virtual mass has been explicitly computed in all directions. In fact, it turned out that of all ellipsoids with a given volume the sphere has the minimum average virtual mass: the conjecture has been verified in an important particular case. I could support the conjecture also by analogy: I succeeded in proving the analogous hydrodynamical minimum property of the circle in two dimensions. So supported, the conjecture deserved to be publicly stated, at least I thought so.

The conjecture is neither proved nor disproved at this date, although G. Szegö and M. Schiffer have found interesting connected results which seem to support it.

18. *A new conjecture.* We consider the plane in which the rectangular coordinates are x and y, and in this plane a region R surrounded by a closed curve C. We seek a function $u = u(x, y)$ that satisfies one or the other of the two boundary conditions

$$(1) \qquad u = 0,$$

$$(2) \qquad \frac{\partial u}{\partial n} = 0$$

along the curve C and the partial differential equation

$$(3) \qquad \frac{\partial^2 u}{\partial x^2} + \frac{\partial^2 u}{\partial y^2} + \nu u = 0$$

within the region R. In (2) n denotes the normal to the curve C; in (3) ν denotes a certain constant. Thus we have two different problems: in the first, we have to solve the differential equation (3) with the boundary condition (1); in the second with (2). Both problems are important in physics in connection with various phenomena of vibration. Both problems have the same trivial solution: $u = 0$ identically. Only for particular values of ν has one or the other problem a non-trivial, that is, not identically vanishing solution u: the problem with the boundary condition (1) for $\nu = \lambda_1, \lambda_2, \lambda_3, \ldots$, the problem with the boundary condition (2) for $\nu = \mu_1, \mu_2, \mu_3, \ldots$. It is

$$0 < \lambda_1 < \lambda_2 \leqq \lambda_3 \leqq \lambda_4 \leqq \ldots,$$

$$0 = \mu_1 < \mu_2 \leqq \mu_3 \leqq \mu_4 \leqq \ldots.$$

These exceptional values $\lambda_1, \lambda_2, \lambda_3, \ldots$ and $\mu_1, \mu_2, \mu_3, \ldots$ are called the *eigenvalues* of the first and the second problem, respectively. The eigenvalues are connected with the frequencies of the characteristic vibrations in the corresponding physical phenomena.

I wish to state a new conjecture: *Let A denote the area of the region R; then, for $n = 1, 2, 3, \ldots$*

$$\mu_n < 4\pi n A^{-1} < \lambda_n.$$

This is a challenging conjecture. As the shape of the region R can arbitrarily vary and n ranges over all the integers 1, 2, 3, . . . , the conjecture covers an immense variety of particular cases of which not too much is known: the conjecture could be exploded by a numerical result concerning any of these particular cases. Yet the conjecture is not too badly supported.

(a) The conjecture is verified for $n = 1, 2, 3, \ldots$ when R is a rectangle. This was, in fact, the particular case that suggested the conjecture.

(b) The conjecture is verified for $n = 1, 2$ when R is of arbitrary shape. This particular case is very different from that mentioned under (a).

(c) The conjecture has been verified by numerical computation in a few particular cases in which the eigenvalues can be explicitly computed: for n up to 25 and for special shapes of R (the circle, a few circular sectors, a few triangles).

(d) It has been known for a long time that

$$\mu_1 \leqq \lambda_1, \quad \mu_2 \leqq \lambda_2, \quad \mu_3 \leqq \lambda_3, \quad \ldots.$$

This agrees with the conjecture.

(e) It is also known that

$$\lim_{n \to \infty} \mu_n n^{-1} = \lim_{n \to \infty} \lambda_n n^{-1} = 4\pi A^{-1}$$

and this also agrees with the conjecture.

Of course, no amount of such verification can prove the conjecture or oblige anybody to believe in it to any degree. Yet such verifications can enhance the interest of the conjecture, spur us on to testing further consequences, and add a zest to scientific investigation which is the *raison d'être* of a conjecture.

Technical details about the conjecture stated (for some of which I am greatly indebted to Mr. Peter Szegö) will be published elsewhere.

19. *Another new conjecture*: "If $F'(x)$ is an algebraic function and all the coefficients $a_1, a_2, a_3, \ldots$ of the series

$$F(x) = a_1x + a_2x^2 + a_3x^3 + \ldots$$

are integers, also $F(x)$ is an algebraic function." In other words: "Unless the integral of an algebraic function is an algebraic function itself, it cannot be represented by a power series with integral coefficients."

For an illustration, consider the expansion

$$\arcsin 2x = 2 \sum_{0}^{\infty} \binom{2n}{n} \frac{x^{2n+1}}{2n+1}$$

The derivative of arc sin $2x$ is an algebraic function and all the coefficients in the expansion of this derivative are integers as we see from the above formula. Yet arc sin $2x$ is not an algebraic function. Therefore, if the proposed conjecture is true, there must be infinitely many coefficients in the above expansion that are not integers. This is easy to verify: if $2n + 1$ is a prime number, it is not a divisor of $2n!$. Thus the case examined supports the conjecture, which is also supported (and has been suggested) by the following fact: if we substitute "rational function" for "algebraic function" in the proposed conjectural statement, we change it into a true and proven statement. The conjecture is also supported by rather vague analogies, by the "atmosphere" surrounding the subject of power series with integral coefficients; see ex. 14.

There are many cases similar to that of arcsin $2x$ that could be easily tested but have not been examined yet. Without a thorough investigation of such accessible consequences such a conjecture should not be printed. I publish it here as an example of an undeveloped, rather imperfectly supported, conjecture.

20. *What is typical?* As far as I can see, there is nothing in the last examples (ex. 12–19) that would disagree with the general impression derived from the foregoing examples. In these last examples, as in the others, the conjecture in question was doubly supported: by some clear facts, and by a "general atmosphere." The former kind of support, by clear facts, seems to me well within the patterns outlined in this chapter and the foregoing: analogy and verified consequences are prominent, merely plausible consequences also play some rôle; support from two very different sides appears as vital. On the whole, ex. 12–19 appear to be typical.

None of the conjectures mentioned in ex. 13–17 has been refuted (to this date, at least). This could look atypical. I have not mentioned here, however, countless other conjectures of mine that have been refuted after a few minutes, or hours, or days; such shortlived conjectures are usually quickly forgotten. Of course, I published only conjectures that passed all the obvious tests I could think of and outlasted at least a few months' work; these are the sturdiest conjectures, with the best chances of survival. Typically, research consists in shaping many conjectures, exploding most of them, and establishing a few.

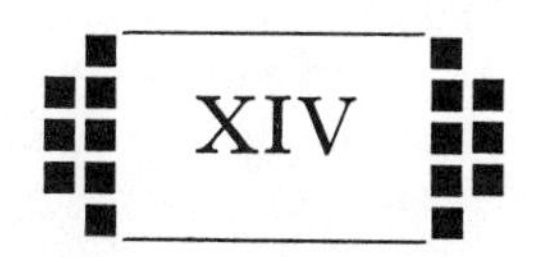

CHANCE, THE EVER-PRESENT RIVAL CONJECTURE

. . . the probability that this coincidence is a mere work of chance is, therefore, considerably less than $(1/2)^{60}$ *. . . Hence this coincidence must be produced by some cause, and a cause can be assigned which affords a perfect explanation of the observed facts.*—G. KIRCHHOFF[1]

1. Random mass phenomena. Everyday speech uses the words "probable," "likely," "plausible," and "credible" in meanings which are not sharply distinguished. Now, we single out the word "probable" and we shall learn to use this word in a specific meaning, as a technical term of a branch of science which is called the "Theory of Probability."[2]

This theory has a great variety of applications and aspects and, therefore, it can be conceived and introduced in various ways. Some authors regard it as a purely mathematical theory, others as a kind or branch of logic, and still others as a part of the study of nature. These various points of view may or may not be incompatible. We have to start by studying one of them, but we should not commit ourselves to any of them. We shall change our position somewhat in the next chapter, but in the present chapter we choose the viewpoint which is the most convenient in the great majority of applications and which the beginner can master most quickly. We regard here the theory of probability as a part of the study of nature, as the theory of certain observable phenomena, the *random mass phenomena*.[3] We can understand pretty clearly what this term means if we compare a few familiar examples of such phenomena.

[1] *Abhandlungen der k. Akademie der Wissenschaften*, Berlin, 1861, p. 79.

[2] In the foregoing, the words "probable" and "probability" have been sometimes used in a non-technical sense, but this will be carefully avoided in the present chapter and the next. The words "likely" and "likelihood" will be introduced as technical terms later in this chapter.

[3] In this essential point, and in several other points, the present exposition follows the views of Richard von Mises although it deviates from his definition of mathematical probability; cf. his book *Probability, Statistics, and Truth.*

(1) *Rainfall.* Rainfall is a mass phenomenon. It consists of a very great number of single events, of the fall of a very great number of raindrops. These raindrops, although very similar to each other, differ in various respects: in size, in the place where they strike the ground, etc. There is something in the behavior of the raindrops that we properly describe as "random." In order to understand clearly the meaning of this term let us imagine an experiment.

Let us observe the first drops on the pavement as the rain starts falling. We observe the pavement in the middle of some large public square, sufficiently far from buildings or trees or anything that could obstruct the rain. We focus our attention on two stones which we call the "right-hand stone" and the "left-hand stone." We observe the drops falling on these stones and we note the order in which they strike. The first drop falls on the left-hand stone, the second drop on the right, the third again right, the fourth left, and so on, without apparent regularity as

$$L\ R\ R\ L\ L\ L\ R\ L\ R\ L\ R\ R\ L\ R\ R$$

(R for right, L for left). There is no regularity in this succession of the raindrops. In fact, having observed a certain number of drops, we cannot reasonably predict which way the next drop will fall. We have noted above fifteen entries. Looking at them, can we predict what the sixteenth entry will be, R or L? Obviously, we cannot. On the other hand, there is some sort of regularity in the succession of the raindrops. In fact, we can confidently predict that at the end of the rain the two stones will be equally wet. That is, the number of drops striking each stone will be very nearly proportional to the area of its free horizontal surface. Nobody doubts that this is so, and the meteorologists certainly assume that this is so in constructing their rain-gauges. Yet there is something paradoxical. We can foresee what will happen in the long run, but we cannot foresee the details. The rainfall is a typical random mass phenomenon, *unpredictable in certain details, predictable in certain numerical proportions of the whole.*

(2) *Boys among the newborn.* In a hospital, the newborn babies are registered in order as they are born. Boys and girls (B and G) follow each other without apparent regularity as

$$G\ B\ B\ G\ B\ G\ B\ B\ G\ G\ B\ B\ B\ G\ G$$

Although we cannot predict the details of this random succession, we can well predict an important feature of the final result obtained by summing up all such registrations in the United States during a year: the number of the boys will be greater than the number of the girls and, in fact, the ratio of these two numbers will be little different from the ratio 51.5 : 48.5. The number of births in the United States is about 3 millions per year. We have here a random mass phenomenon of considerable dimensions.

(3) *A game of chance.* We toss a penny repeatedly, noting each time which side it shows, "heads" or "tails" (H or T). We obtain so a succession without apparent regularity as

$$T\ H\ H\ H\ T\ H\ T\ H\ H\ T\ H\ T\ H\ T\ T$$

If we have the patience to toss the penny a few hundred times, a definite ratio of heads to tails emerges, which does not change much if we prolong our experiment still further. If our penny is "unbiased," the ratio 50 : 50 of heads to tails should appear in the long run. If the penny is biased, some other ratio will come into view. At any rate we see again the characteristic features of a random mass phenomenon. Constant proportions emerge in the long run, although the details are unpredictable. There is a certain aggregate regularity, in spite of the irregularity of the individual happenings.

2. The concept of probability. In the year 1943 the number of births in the United States, male, female, and total, was

$$1{,}506{,}959 \qquad 1{,}427{,}901 \qquad 2{,}934{,}860,$$

respectively. We call

1,506,959 the frequency of the male births,

1,427,901 the frequency of the female births.

We call

$$\frac{1{,}506{,}959}{2{,}934{,}860} = 0.5135$$

the relative frequency of the male births and

$$\frac{1{,}427{,}901}{2{,}934{,}860} = 0.4865$$

the relative frequency of the female births. In general, if an event of a certain kind occurs in m cases out of n, we call m the *frequency* of occurrence of that kind of event and m/n its *relative frequency*.

Let us imagine that, throughout the whole year, the births are successively registered in the whole United States (as in the hospital that we have mentioned in the foregoing section). If we look at the succession of male and female births, we have before us an extremely long series of almost three million entries beginning like

$$G\ B\ B\ G\ B\ G\ B\ B\ G\ G\ B\ B\ B\ G\ G.$$

As the mass phenomenon unfolds, we have, at each stage of the observation, a certain frequency of male births and also a certain relative frequency. Let us note, after 1, 2, 3, . . . observations, the frequencies and the relative frequencies found up to that point:

Observations	*Event*	*Frequency of B*	*Relative frequency*
1	G	0	$0/1 = 0.000$
2	B	1	$1/2 = 0.500$
3	B	2	$2/3 = 0.667$
4	G	2	$2/4 = 0.500$
5	B	3	$3/5 = 0.600$
6	G	3	$3/6 = 0.500$
7	B	4	$4/7 = 0.571$
8	B	5	$5/8 = 0.625$
9	G	5	$5/9 = 0.556$
10	G	5	$5/10 = 0.500$
11	B	6	$6/11 = 0.545$
12	B	7	$7/12 = 0.583$
13	B	8	$8/13 = 0.615$
14	G	8	$8/14 = 0.571$
15	G	8	$8/15 = 0.533$

As far as we have tabulated it, the relative frequency oscillates pretty strongly (between the limits 0.000 and 0.667). Yet we have here only a very small number of observations. As we go further and further, the oscillations of the relative frequency will become less and less violent, and we can confidently expect that in the end it will oscillate very little about its final value 0.5135. *As the number of observations increases, the relative frequency appears to settle down to a stable final value, in spite of all the unpredictable irregularities of detail.* Such behavior, the emergence of a stable relative frequency in the long run, is typical of random mass phenomena.

An important aim of any theory of such phenomena must be to predict the final stable relative frequency or *long range relative frequency.* We have to consider the *theoretical value of long range relative frequency* and we shall call this theoretical value *probability.*

We wish to clarify this concept of probability. Naturally, we begin with the study of mass phenomena for which we can predict the long range relative frequency with some degree of reasonable confidence.

(1) *Balls in a bag.* A bag contains p balls of various colors among which there are exactly f white balls. We use this simple apparatus to produce a random mass phenomenon. We draw a ball, we look at its color and we write W if the ball is white, but we write D if it is of a different color. We put back the ball just drawn into the bag, we shuffle the balls in the bag, then

we draw again one and note the color of this second ball, W or D. In proceeding so, we obtain a random sequence similar to those considered in sect. 1:

$$W\ D\ D\ D\ W\ D\ D\ W\ W\ D\ D\ D\ W\ W\ D.$$

What is the long range relative frequency of the white balls?

Let us discuss the circumstances in which we can predict the desired frequency with reasonable confidence. Let us assume that the balls are homogeneous and exactly spherical, made of the same material and having the same radius. Their surfaces are equally smooth, and their different coloration influences only negligibly their mechanical behavior, if it has any influence at all. The person who draws the balls is blindfolded or prevented in some other manner from seeing the balls. The position of the balls in the bag varies from one drawing to the other, is unpredictable, beyond our control. Yet the permanent circumstances are well under control: the balls are all the same shape, size, and weight; they are *undistinguishable* by the person who draws them.

Under such circumstances we see no reason why one ball should be preferred to another and we naturally expect that, in the long run, each ball will be drawn approximately *equally often*. Let us say that we have the patience to make 10,000 drawings. Then we should expect that each of the p balls will appear about

$$\frac{10{,}000}{p} \text{ times.}$$

There are f white balls. Therefore, in 10,000 drawings, we expect to get white

$$f\,\frac{10{,}000}{p} = 10{,}000\,\frac{f}{p} \text{ times;}$$

this is the expected frequency of the white balls. To obtain the relative frequency, we have to divide the frequency by the number of observations, or drawings, that is, by 10,000. And so we are led to the statement: the long range relative frequency, or *probability*, of the white balls is f/p.

The letters f and p are chosen to conform to the traditional mode of expression. As we have to draw one of the p balls, we have to choose one of p *possible* cases. We have good reasons (equal condition of the p balls) not to prefer any of these p possible cases to any other. If we wish that a white ball should be drawn (for example, if we are betting on white), the f white balls appear to us as *favorable* cases. Hence we can describe the probability f/p as the *ratio of the number of favorable cases to the number of possible cases.*

Pulling a ball from a bag, putting the ball back into the bag, shaking the bag, pulling another ball, and repeating this n times seems to be a pretty silly occupation. Do we waste our time in studying such a primitive game?

I do not think so. The bag and the balls, handled in the described manner, generate a random mass phenomenon which is particularly simple and accessible. Generalization naturally starts from the simplest, the most transparent particular case. The science of dynamics was born when Galileo began studying the fall of heavy bodies. The science of probability was born when Fermat and Pascal began studying games of chance which depend on casting a die, or drawing a card from a pack, or drawing a ball from a bag. The fundamental concepts and laws of dynamics can be extracted from the simple phenomenon of falling bodies. We use the bag and the balls to understand the fundamental concept of probability.

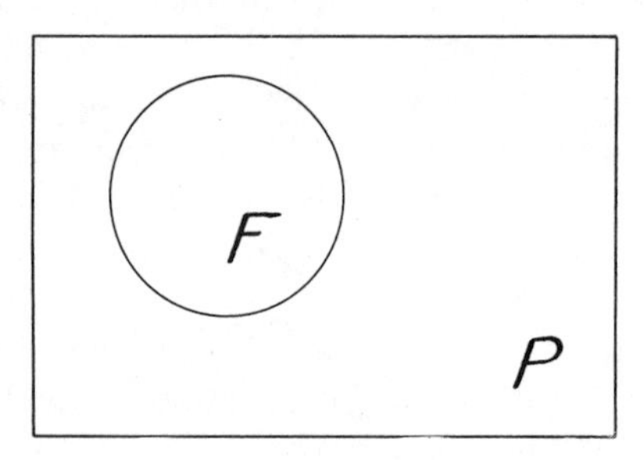

Fig. 14.1. Probability defined by rainfall.

(2) *Rainfall.* We return to the consideration of the random mass phenomenon from which we started in sect. 1. The area of a horizontal surface is P and the area of a certain portion of this surface is F; see fig. 14.1. We observe the raindrops falling on this area P and we are interested in the frequency of the raindrops falling on the subarea F. We are inclined to predict without hesitation the long range relative frequency: the fraction of the total rain over the area that falls on the subarea will be very nearly F/P if the rain consists of more than a few drops. In other words, the probability that a raindrop striking the surface of area P should strike the portion of area F is F/P. If we idealize the rainfall and consider a raindrop as a geometric point, we can also say: the probability that a point falling in the area P should fall in the subarea F is F/P.

In the last statement we consider each point of the area P as a possible case and each point of the subarea F as a favorable case. The number of favorable cases as that of the possible cases is infinite, and it would not make sense to talk about the ratio of infinite numbers. We can consider, however, the area of a surface as the *measure* of the points contained in the surface. Using this term, we can describe the probability F/P as the *ratio of the measure of favorable cases to the measure of possible cases.*

3. Using the bag and the balls. In deriving the fundamental principle of statics, Lagrange replaced an arbitrary system of forces by a suitable system of pulleys. In the light of this Lagrangeian argument (the details of which are not needed here[4]) any case of equilibrium appears as a suitable combination of correctly balanced pulleys. The Calculus of Probability can be viewed in a similar manner; in fact, such a view is suggested by the early history of this science. Seen from this standpoint, any problem of probability appears comparable to a suitable problem about bags containing balls, and any random mass phenomenon appears as similar in certain essential respects to successive drawings of balls from a system of suitably combined bags. Let us illustrate this by a few simple examples.

(1) Instead of tossing a fair penny, we can draw a ball from a bag containing just two balls, one of which is marked with an *H* and the other with a *T* (heads and tails). Instead of casting an unbiased die, we can draw a ball from a bag containing exactly six balls, marked with 1, 2, 3, 4, 5, or 6 spots, respectively. Instead of drawing a card from a pack of cards, we can draw a ball from a bag containing 52 balls, suitably marked. It seems to be intuitively clear that substituting a bag with balls for pennies, dice, cards, and other similar contrivances in a suitable way, we do not change the odds in the usual games of chance. At least, we do not change the chances in that idealized version of these games in which the contrivances used (pennies, dice, etc.) are supposed to be perfectly symmetrical and, correspondingly, certain fundamental chances perfectly equal.

(2) Wishing to study the randomness in the distribution of boys and girls among the newborn, we may substitute for the actual mass phenomenon successive drawings from a bag containing 1,000 balls, 515 marked with *B* and 485 marked with *G*. This substitution is, of course, theoretical and, as every theory is bound to be, it is tentative and approximative. Yet the point is that the bag and the balls enable us to formulate a theory.

(3) A meteorologist registers the succession of rainy and rainless days in a certain locality. His observations seem to show that, on the whole, each day tends to resemble the foregoing day: rainless days seem to follow rainless days more easily than rainy days and, similarly, rainy days seem to follow rainy days more easily than rainless days. Of course, a dependable regularity appears only in a long series of observations; the details are irregular, seem to be random.

The meteorologist may wish to express more clearly his impressions that we have just sketched. If he wishes to formulate a theory in terms of probability, he may consider three bags. Each bag contains the same number of balls, let us say 1,000 balls. Some of the balls are white, the others are black (white for rainless, black for rainy). Yet there are important differences between the bags. Each bag bears an inscription, easily visible to the person who draws the balls. One bag is inscribed "START," another "AFTER WHITE," and the third "AFTER BLACK." The ratio of balls of different color is different in different bags. In each bag the ratio of white balls to

[4] See E. Mach, *Die Mechanik*, p. 59–62.

black balls approximates the observable ratio of rainless days to rainy days, but in different circumstances. In the bag "START" the ratio is that of rainless days to rainy days throughout the year, in the bag "AFTER WHITE" the ratio is that of rainless days to rainy days following a rainless day, and in the bag "AFTER BLACK" the ratio is that of rainless days to rainy days following a rainy day. Therefore, the bag "AFTER WHITE" contains *more* white balls than the bag "AFTER BLACK." The balls are drawn successively and each ball drawn, when its color has been noticed, is replaced into the bag from which it was drawn. The bag "START" is used but once, for the first ball. If the first ball is white, we use the bag "AFTER WHITE" for the second ball, but if the first ball is black, the second ball is drawn from the bag "AFTER BLACK." And so on, the color of the ball just drawn determines the bag from which the next ball should be drawn.

It is just a theory that the succession of white and black balls drawn under the described circumstances imitates the succession of rainless and rainy days with a reasonable approximation. Yet, on the face of it, this theory does not seem to be out of place. At any rate, this theory, or some similar theory, could deserve to be confronted with the observations.

(4) Take any English text (from Shakespeare, if you prefer) and replace each of the letters *a*, *e*, *i*, *o*, *u*, and *y* by V and each of the remaining twenty letters by C. (V means vowel and C means consonant.) You obtain a pattern as

$$C\ V\ C\ V\ V\ C\ C\ V\ C\ C\ V\ C\ V\ C\ C\ .$$

This irregular sequence is in some way opposite to that discussed in the foregoing subsection (3): each day tends to be like the foregoing day, but each letter tends to be unlike the foregoing letter. Still, we could imitate the succession of vowels and consonants by a succession of white and black balls drawn from three bags bearing the same inscriptions as before (in subsection (3)), yet the ratio of white balls to black balls should not be the same as before. To imitate realistically the succession of vowels and consonants the bag "AFTER WHITE" should contain *less* white balls than the bag "AFTER BLACK."

(5) There are two bags. The first bag contains p balls among which there are f white balls. The second bag contains P chips among which there are F white chips. Using both hands, I draw from both bags at the same time, a ball with the left hand and a chip with the right hand. What is the probability that both the ball and the chip turn out to be white?

We could, of course, repeat this primitive experiment sufficiently often, perhaps a thousand times, and so obtain an approximate value for the desired probability. Yet we can also try to guess it, and that is more interesting.

The result of the two simultaneous drawings is a "couple," consisting of a ball and a chip. There are p balls and P chips. As any ball can be coupled with any chip, there are pP possible couples; they are shown in fig. 14.2 where $p = 5, f = 2, P = 4, F = 3$. There is no reason to prefer any of the p balls to any other ball, or any of the P chips to any other chip. There seems to be *no reason to prefer any of the pP couples to any other couple.* In fact, in performing the experiment with the two bags, I am supposed to proceed blindly, at random, so that each hand draws independently of the other. "Let not thy left hand know what thy right hand doeth." It seems incredible that the chances of the ball that I draw with my left hand should be influenced by the chip that I draw with my right hand. Why should ball no. 1 be any more attracted by chip no. 1 than it is by chip no. 2?

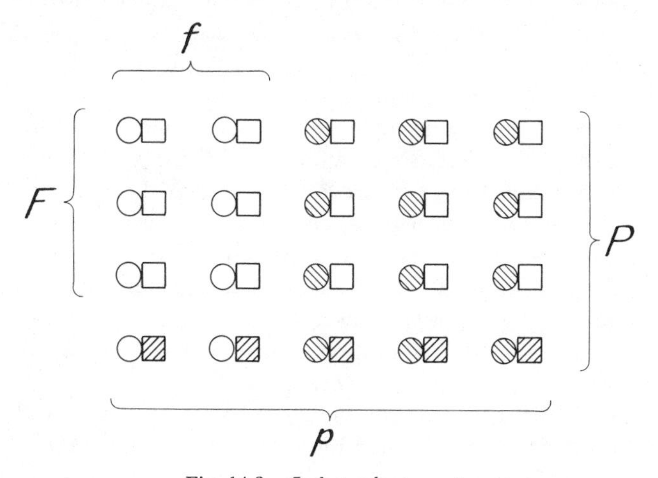

Fig. 14.2. Independent events.

And so we can imagine a bag, containing pP mechanically indistinguishable objects (each object is a couple, a ball attached to a chip); one drawing from this one bag appears *equivalent* to the two simultaneous drawings from the two bags described at the outset. We have so pP possible cases; it remains to find the number of favorable cases. A glance at fig. 14.2 shows that there are fF couples consisting of a white ball and a white chip. And so we obtain the value of the desired probability: it is

$$\frac{fF}{pP} = \frac{f}{p} \cdot \frac{F}{P}$$

the *product* of two probabilities. In fact, f/p is the probability of drawing a white ball from the first bag, and F/P the probability of drawing a white chip from the second bag.

The essential point about the ball and the chip is that the drawing of one does not influence the chances of the other. In the usual terminology of the calculus of probability, such events are called *independent* of each other; the joint happening of both events is viewed as a *compound* event. The foregoing consideration motivates the rule: *The probability of a compound event is the product of the probabilities of the constituent events, provided that these constituent events are mutually independent.*

4. The calculus of probability. Statistical hypotheses. The theory of probability, as we see it, is a part of the study of nature, the theory of random mass phenomena.

The most striking achievement of the physical sciences is prediction. The astronomers predict with precision the eclipses of the sun and the moon, the position of the planets, and the return of comets which evade observation for several years. A great astronomer (Leverrier) succeeded even in predicting the position of a planet (Neptune), the very existence of which was not known before. The theory of probability predicts the frequencies in certain mass phenomena with some amount of success.

The astronomers base their predictions on former observations, on the laws of mechanics, the law of gravitation, and long difficult computations. Any branch of physical science bases its predictions on some theory or, we can say, on some conjecture, since no theory is certain and so every theory is a more or less reasonable, more or less well-supported, conjecture. In trying to predict the frequencies in a certain random mass phenomenon from the theory of probability we have to make some theoretical assumption about the phenomenon. Such an assumption, which has to be expressed in terms of probability concepts, is called a *statistical hypothesis*.

When we apply the theory of probability we have to compute probabilities (which are theoretical, approximate values of relative frequencies). When we try to find a probability, we have a problem to solve. The unknown of this problem is the desired probability. Yet, in order to determine this unknown, we need data and conditions in our problem. The data are usually probabilities and the conditions, on which the relation of the unknown probability to the given probabilities depends, constitute a statistical hypothesis.

As in the applications of the theory of probability the computation of probabilities plays a prominent role, this theory is usually called the *calculus of probability*. Thus, the aim of the calculus of probability is to compute new probabilities on the basis of given probabilities and given statistical hypotheses.

The reader who wishes to peruse the remaining part of this chapter must either know the elements of the calculus of probability, or he must take for granted certain results derived from these elements. Most of the time, the text will state the results without derivation; derivations will be given subsequently, in the First Part of the Exercises and Comments following this chapter, and in the corresponding Solutions. Yet even if the reader does not check the derivation of the results, he ought to have some insight into the underlying theoretical assumptions. We can make such assumptions intuitively understandable: we compare the random mass phenomenon that we examine to drawings from suitably filled bags under suitable conditions, as in the foregoing sect. 3.

The applications of the calculus of probability are of unending variety. The following sections of this chapter attempt to illustrate the principal types of applications by suitable elementary examples. Stress will be laid on the motivation of these applications, that is, on such preliminary considerations as make the choice of procedure plausible.

5. Straightforward prediction of frequencies. At the beginning of its history the calculus of probability was essentially a theory of certain games of chance. Yet the predictions of this theory were not tested experimentally on a large scale until modern times. We begin by discussing an experiment of this kind.

(1) W. F. R. Weldon cast 12 dice 26,306 times, noting each time how many of these 12 dice have shown more than four spots.[5] The results of his observations are listed in column (4) of Table I; column (1) shows the number of the dice among the 12 that have turned up five or six spots. Thus, in 26,306 trials it never happened that all twelve dice showed more than four spots. The most frequent case was that in which four out of the twelve dice showed five or six spots; this happened 6,114 times.

Table I

(1) Nr. of 5 or 6	(2) Excess I	(3) Predicted I	(4) Observed	(5) Predicted II	(6) Excess II
0	+ 18	203	185	187	+ 2
1	+ 67	1216	1149	1146	− 3
2	+ 80	3345	3265	3215	− 50
3	+ 101	5576	5475	5465	− 10
4	+ 159	6273	6114	6269	+ 155
5	− 176	5018	5194	5115	− 79
6	− 140	2927	3067	3043	− 24
7	− 76	1255	1331	1330	− 1
8	− 11	392	403	424	+ 21
9	− 18	87	105	96	− 9
10	− 1	13	14	15	+ 1
11	− 3	1	4	1	− 3
12	0	0	0	0	0
Total	0	26,306	26,306	26,306	0

How can the theory predict the observed numbers listed in column (4) of Table I? If we assume that the dice are "fair" and that the trials with different dice, or with the same die at different times, are independent of

[5] *Philosophical Magazine*, ser. 5, vol. 50, 1900, p. 167–169; in a paper by Karl Pearson.

each other, we can compute the relevant probabilities. Under our assumption (which is properly termed a "statistical hypothesis") the probability that exactly 4 dice out of 12 should show 5 or 6 spots is

$$P = 495 \left(\frac{1}{3}\right)^4 \left(\frac{2}{3}\right)^8 = \frac{126,720}{531,441}.$$

Now, by definition, the probability is the theoretical value of long range relative frequency. If the event with probability P shows itself m times in n trials, we expect that

$$\frac{m}{n} = P \text{ approximately}$$

or

$$m = Pn \text{ approximately.}$$

Therefore, we should expect that exactly 4 dice will show five or six spots out of the 12 dice cast in about

$$Pn = \frac{126,720}{531,441} 26,306 = 6,273$$

cases out of $n = 26,306$ trials. (Observe that we can compute this number 6,273 before the trials start.) Now, this predicted value 6,273 does not seem to be "very different" from the observed number 6,114, and so our first impression about the practical applicability of the theory of probability may be quite good.

The number 6,273 is listed in column (3) of Table I at the proper place, in the same row as the number 4 in column (1). All the numbers in column (3) are similarly computed. In order to compare more conveniently the predicted values in column (3) with the observed numbers in column (4), we list the differences (predicted less observed) in column (2). With their meaning in mind, we survey the columns (2), (3), and (4). Is the agreement between experience and theory satisfactory? Are the observed numbers sufficiently close to the predicted values?

There is, obviously, some agreement between the columns (3) and (4). Both columns of numbers have the same general aspect: the maximum is attained at the same point (in the same row) and the numbers first increase to the maximum and then decrease steadily to 0 in very much the same fashion in both columns. The deviation of the observed number from the predicted value appears relatively small in most cases; the agreement, at a first glance, looks quite good. On the other hand, however, the number of trials, 26,306, appears pretty large. Are the deviations sufficiently small in view of the large number of trials?

This seems to be the right question. Yet we cannot answer it off-hand; we had better postpone it till we know a little more; see sect. 7 (3). Yet without any special knowledge, just with a little common sense, we can draw quite a sharp conclusion from Table I. A physicist would easily notice the following point about the columns (3) and (4). The differences are listed in column (2). Some of these differences are positive, others negative. If these differences were randomly distributed, the + and — signs should be intermingled in some disorderly fashion. In fact, however, the + and — signs are sharply separated: the theoretical values are too large up to a certain point, and too small from that point onward. In such a case, the physicist speaks of a *systematic* deviation of the theory from the experiment, and he regards such a systematic deviation as a grave objection against the theory.

And so the agreement between the theory of probability and Weldon's observations, which looked quite good at first, begins to look much less good.

(2) Yet who is responsible for that systematic deviation? The theoretical values have been computed according to the rules of the calculus of probability on the basis of a certain assumption, a "statistical hypothesis." We need not blame the rules of the calculus; the fault may be with the statistical hypothesis. In fact, this statistical hypothesis has a weak point: we assumed that the dice used in the experiment were "fair." When gentlemen play a game of dice, they should assume that the dice are fair, but for a naturalist such an assumption is unwarranted.

In fact, let us look at the example of the physicist. Galileo discovered the law of falling bodies that we write today in the usual notation as an equation:

$$s = gt^2/2;$$

s stands for space (distance), t for time. More exactly, Galileo discovered the *form* of the dependence of s on t: the distance is proportional to the square of the time t. Yet he made no theoretical prediction about the constant g that enters into this proportionality; the suitable value of g has to be found by experiments. In this respect, as in many other respects, natural science followed the example of Galileo; in countless cases the theory yielded the general form of a natural law, and the experiment had to determine the numerical values of the constants that enter into the mathematical expression of the law. And this procedure works in our example, too.

If a die is "fair," none of the six faces is preferable to the others, and so the probability for casting 5 or 6 spots is

$$\frac{2}{6} = \frac{1}{3}.$$

Even if the die is not fair, there is a certain probability p for casting 5 or 6 spots; p may be different from $1/3$. (Yet not very different in an ordinary die, otherwise we would consider the die as "loaded.") We take p as a constant that has to be determined by experiment. And now, we modify our original statistical hypothesis: we *assume* that all twelve dice used have the *same probability* p for showing 5 or 6 spots. (This is a simple assumption but, of course, pretty arbitrary. We cannot believe that it is exactly true; we can only hope that it is not very far from the truth. There is virtually no chance that the dice are exactly equal, but they may be only slightly different.) We keep unchanged the other part of our former statistical hypothesis (different dice and different trials are considered as independent).

On the basis of this new statistical hypothesis we can again assign theoretical values corresponding to the observations listed in column (4) of Table I. For example, the theoretical value corresponding to the observed value 6,114 is

$$495\, p^4 (1 - p)^8\, 26{,}306;$$

it depends on p, and also the theoretical values corresponding to the other numbers in column (4) depend on p.

It remains to determine p from the experiments that we are examining. We cannot hope to determine p from experiments exactly, only in some reasonable approximation. If we change our standpoint for a moment and consider the casting of a single die as a trial,

$$12 \times 26{,}306 = 315{,}672$$

trials have been performed; this is a very large number. The frequency of the event "five or six spots" can be easily derived from the column (4) of Table I. We find as the value of the relative frequency

$$\frac{106{,}602}{315{,}672} = 0.3376986;$$

we take this relative frequency, resulting from a very large number of trials, for the value of p. (We assume so for p a value slightly higher than 1/3.)

Once p is chosen, we can compute theoretical values corresponding to the observed frequencies. These theoretical values are tabulated in column (5) of Table I. Thus the columns (3) and (5) give theoretical values corresponding to the same observed numbers, but computed under different statistical hypotheses. In fact, the two statistical hypotheses differ only in the value of p; column (3) uses $p = 1/3$, column (5) uses the slightly higher value derived from the observations. (Column (3) can be computed before the observations, but column (5) cannot.) The differences between corresponding items of columns (5) and (4) are listed in column (6).

There is little doubt that the theoretical values in column (5) fit the observations much better than those in column (3). In absolute value, the differences in column (6) are, with just one exception, less than, or equal to, the differences in column (2) (equal in just three cases, much less in most cases). In opposition to column (2), the signs + and − are intermingled in column (6), so that they yield no ground to suspect a systematic deviation of the theoretical values in column (5) from the experimental data in column (4).

(3) Judged by the foregoing example, the theory of probability seems to be quite suitable for describing mass phenomena generated by such gambling devices as dice. If it were not suitable for anything else, it would not deserve too much attention. Let us, therefore, consider one more example.

As reported by the careful official Swiss statistical service, there were exactly 300 deliveries of triplets in Switzerland in the 30 years from 1871 to 1900. (That is, 900 triplets were born. In talking of deliveries, we count the mothers, not the babies.) The number of all deliveries (some of triplets, some of twins, most of them, of course, of just one child) during the same period in the same geographical unit was 2,612,246. Thus, we have here a mass phenomenon of considerable proportions, but the event considered, the birth of triplets, is a *rare event*. The average number of deliveries per year is

$$2{,}612{,}246/30 = 87{,}075,$$

the average number of deliveries of triplets only

$$300/30 = 10.$$

Of course, the event happened more often in some years, in others less often than the average 10, and in some years exactly 10 times. Table II gives

Table II

Triplets born in Switzerland 1871–1900.

(1) Deliveries	(2) Years obs.	(3) Years theor.	(4) (2) cumul.	(5) (3) cumul.
0	0	0.00	0	0.00
1	0	0.00	0	0.00
2	0	0.09	0	0.09
3	1	0.21	1	0.30
4	0	0.57	1	0.87
5	1	1.14	2	2.01
6	1	1.89	3	3.90
7	5	2.70	8	6.60
8	1	3.39	9	9.99
9	4	3.75	13	13.74
10	4	3.75	17	17.49
11	4	3.42	21	20.91
12	3	2.85	24	23.76
13	2	2.16	26	25.92
14	1	1.59	27	27.51
15	2	1.02	29	28.53
16	0	0.66	29	29.19
17	1	0.39	30	29.58
18	0	0.21	30	29.79
19	0	0.12	30	29.91

the relevant details in column (2). We see there (in the row that has 10 in the first column) that there were in the period considered exactly 4 years in which exactly 10 deliveries of triplets took place. As the same column (2) shows, no year in the period had less than 3 such deliveries, none had more than 17, and each of these extreme numbers, 3 and 17, turned up in just one year.

The numbers of column (2) seem to be dispersed in some haphazard manner. It is interesting to note that the calculus of probability is able to match the irregular looking observed numbers in column (2) by theoretical numbers following a simple law; see column (3). The agreement of columns (2) and (3), judged by inspection, does not seem to be bad; the difference between the two numbers, the observed and the theoretical, is less than 1 in absolute value, except in two cases. Yet in these two cases (the rows with 7 and 8 in the first column) the difference is greater than 2 in absolute value.

There is a device that allows us to judge a little better the agreement of the two series of numbers. The column (4) of Table II contains the numbers of column (2) "cumulatively." For example, consider the row that has 7 in column (1); it has 5 in column (2) and 8 in column (4). Now

$$8 = 0 + 0 + 0 + 1 + 0 + 1 + 1 + 5;$$

that is, 8 is the sum, or the "accumulation", of all numbers in column (2) up to the number 5, inclusively, in the respective row. (In other words, 8 is the number of those years of the period in which the number of deliveries of triplets did *not exceed* 7.) Column (5) contains the numbers of column (3) "cumulatively", and so the columns (4) and (5) are analogously derived from the observed numbers in column (2) and the theoretical numbers in column (3), respectively. The agreement between columns (4) and (5) looks excellent; the difference is less than 1 in absolute value except in just one case, where it is still less than 2.

6. Explanation of phenomena. Ideas connected with the concept of probability play a rôle in the explanation of phenomena, and that is true of phenomena dealt with by any science, from physics to the social sciences. We consider two examples.

(1) Gregor Mendel (1822–1884), experimenting with the cross-breeding of plants, became the founder of a new science, genetics. Mendel was, by the way, an abbot in Moravia, and carried out his experiments in the garden of his monastery. His discovery, although very important, is very simple. To understand it we need only the description of one experiment and an intuitive notion of probability. To make things still easier, we shall not discuss one of Mendel's own experiments, but an experiment carried out by one of his followers.[6]

Of two closely related plants (different species of the same genus) one has

[6] By Correns; see W. Johannsen, *Elemente der exakten Erblichkeitslehre*, Jena 1909, p. 371.

white flowers and the other rather dark red flowers. The two plants are so closely related that they can fertilize each other. The seeds resulting from such crossing develop into hybrid plants which have an intermediate character: the hybrids have pink flowers. (In fig. 14.3 red is indicated by more, pink by less, shading.) If the hybrid plants are allowed to become self-fertilized, the resulting seeds develop into a third generation of plants in which all three kinds are represented: there are plants with white, plants with pink, and plants with red flowers. Fig. 14.3 represents schematically the relations between the three subsequent generations.

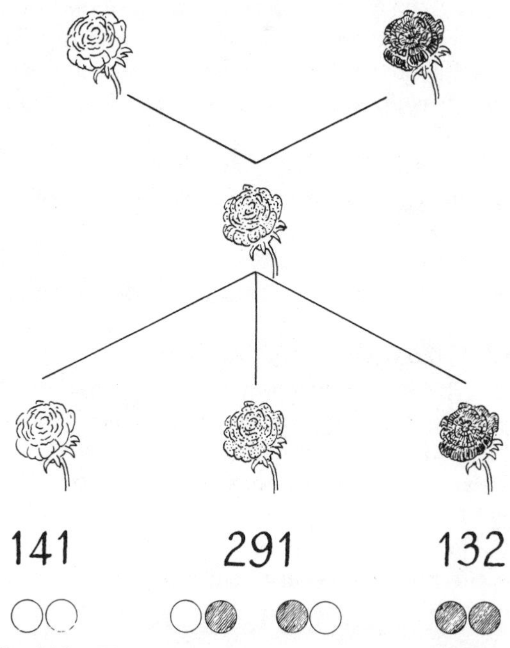

Fig. 14.3. Three generations in a Mendelian experiment.

Yet the most striking feature of the phenomenon is the numerical proportion in which the three different kinds of plants of the third generation are produced. In the experiment described, 564 plants of the third generation have been observed. Among them, those two kinds of plants that resemble one or the other grandparental plant were about equally numerous: there were 141 plants with white flowers and 132 plants with red flowers in the third generation. Yet the plants resembling the hybrid parental plants were more numerous: there were 291 plants with pink flowers in the third generation. We can conveniently survey these numbers in fig. 14.3. We easily notice that these numbers given by the experiment are approximately in a simple proportion:

$$141 : 291 : 132 \text{ almost as } 1 : 2 : 1.$$

This simple proportion invites a simple explanation.

Let us begin at the beginning. The experiment began with the crossing of two different kinds of plants. Any flowering plant arises from the union of two reproductive cells (an ovule and a grain of pollen). The pink-flowering hybrids of the second generation arose from two reproductive cells of different extraction. As the pink-flowering plants of the third generation are similar to those of the second generation, it is natural to assume that they were similarly produced, by two reproductive cells of *different* kinds. This leads us to suppose that the pink-flowering hybrids of the second generation *have* two different kinds of reproductive cells. Supposing this, however, we may perceive a possibility of explaining the mixed offspring. In fact, let us see more clearly what would happen if the pink-flowering hybrids of the second generation actually *had* two different kinds of reproductive cells, which we may call "white" and "red" cells. When two such cells are combined, the combination can be white with white, or red with red, or one color with the other, and these three different combinations *could* explain the three different kinds of plants in the third generation; see fig. 14.3.

After this remark, it should not be difficult to explain the numerical proportions. The deviation of the actually observed proportion 141 : 291 : 132 from the simple proportion 1 : 2 : 1 appears as random. That is, it looks like the deviation of observed frequencies from underlying probabilities. This leads us to wondering what the probabilities of the two kinds of cells are, or in which proportion the "white" and "red" cells are produced. As there are about as many white-flowering as red-flowering plants in the third generation, we can hardly refrain from trying the simplest thing: let us assume that the "white" and "red" reproductive cells are produced in equal numbers by the pink-flowering plants. Finally, we are almost driven to compare the random encounter of two reproductive cells with the random drawing of two balls, and so we arrive at the following simple problem.

There are two bags containing white and red balls, and no balls of any other color. Each bag contains just as many white balls as red balls. With both hands, I draw from both bags, one ball from each. Find the probability for drawing two white balls, two balls of different colors, and two red balls.

As it is easily seen (cf. sect. 3 (5)), the required probabilities are

$$\frac{1}{4}, \quad \frac{2}{4}, \quad \frac{1}{4},$$

respectively. We perceive now a simple reason for the proportion 1 : 2 : 1 that seems to underlie the observed numbers, and so doing we come very close to Mendel's essential concepts.

(2) The concept of random mass phenomena plays an important rôle in physics. In order to illustrate this rôle, we consider the velocity of chemical reactions.

Relatively crude observations are sufficient to suggest that the speed of a chemical change depends on the concentration of the reacting substances. (By concentration of a substance we mean its amount in unit volume.) This dependence of the chemical reaction velocity on the concentration of the reactants was soon recognized, but the discovery of the mathematical form of the dependence came much later. An important particular case was noticed by Wilhelmy in 1850, and the general law was discovered by two Norwegian chemists, Guldberg and Waage, in 1867. We now outline, in a particular case and as simply as we can, some of the considerations that led Guldberg and Waage to their discovery.

We consider a bimolecular reaction. That is, two different substances, A and B, participate in the reaction which consists in the combination of one molecule of the first substance A with one molecule of the second substance B. The substances A and B are dissolved in water, and the chemical change takes place in this solution. The substances resulting from the reaction do not participate further in the chemical action; they are inactive in one way or another. For example, they may be insoluble in water and deposited in solid form.

The solution in which the reaction takes place consists of a very great number of molecules. According to the ideas of the physicists (the kinetic theory of matter) these molecules are in violent motion, traveling at various speeds, some at very high speed, and colliding now and then. If a molecule A collides with a molecule B, the two may get so involved that they exchange some of their atoms: the chemical reaction in which we are interested consists of such an exchange, we imagine. Perhaps it is necessary for such an exchange that the molecules should collide at a very high speed, or that they should be disposed in a favorable position with respect to each other in the moment of their collision. At any rate, the more often it happens that a molecule A collides with a molecule B, the more chance there is for the chemical combination of two such molecules, and the higher the velocity of the chemical reaction will be. And so we are led to the conjecture: *the reaction velocity is proportional to the number of collisions between molecules A and molecules B.*

We could not predict exactly the number of such collisions. We have before us a random mass phenomenon like rainfall. Remember fig. 14.2; there, too, we could not predict exactly how many raindrops would strike the subarea F. Yet we could predict that the number of raindrops striking the subarea F would be *proportional* to the number of raindrops falling on the whole area P. (The proportionality is approximate, and the factor of proportionality is F/P, as discussed toward the end of sect. 2.) Similarly, we can predict that the number of collisions in which we are interested (between any molecule A and any molecule B) will be proportional to the number of the molecules A. Of course, it will also be proportional to the number of the molecules B, and so finally proportional to the *product* of these two numbers. Yet the number of the molecules of a substance is proportional to the concentration of that substance, and so our conjecture leads us to the following statement: *the reaction velocity is proportional to the product of the concentrations.*

We arrived at a particular case of the general law of chemical mass action discovered by Guldberg and Waage. This is the particular case appropriate for the particular circumstances considered. On the basis of the law of mass action it is possible to compute the concentration of the reacting substances at any given moment and to predict the whole course of the reaction.

7. Judging statistical hypotheses. We start from an anecdote.[7]

(1) "One day in Naples the reverend Galiani saw a man from the Basilicata who, shaking three dice in a cup, wagered to throw three sixes; and, in fact, he got three sixes right away. Such luck is possible, you say. Yet the man succeeded a second time, and the bet was repeated. He put back the dice in the cup, three, four, five times, and each time he produced three sixes. 'Sangue di Bacco,' exclaimed the reverend, 'the dice are loaded!' And they were. Yet why did the reverend use profane language?"

The reverend Galiani drew a plausible conclusion of a very important type. If he discovered for himself this important type of plausible inference on the spur of the moment, his excitement is quite understandable and I, personally, would not reproach him for his mildly profane language.

The correct thing is to treat everybody as a gentleman until there is some definite evidence to the contrary. Quite similarly, the correct thing is to engage in a game of chance under the assumption that it is fairly played. I do not doubt that the reverend did the correct thing and assumed in the beginning that that man from the Basilicata had fair dice and used them fairly. Such an assumption, correctly stated in terms of probability, is a statistical hypothesis. A statistical hypothesis generally assumes the values of certain probabilities. Thus, the reverend assumed in the beginning, more or less explicitly, that any of the dice involved will show six spots with the probability 1/6. (We have here exactly the same statistical hypothesis as in sect. 5 (1).)

The calculus of probability enables us to compute desired probabilities from given probabilities, on the basis of a given statistical hypothesis. Thus, on the basis of the statistical hypothesis adopted by the reverend at the beginning, we can compute the probability for casting three sixes with three dice; it is

$$(1/6)^3 = 1/216,$$

a pretty small probability. The probability for repeating this feat twice,

[7] J. Bertrand, *Calcul des probabilités*, p. VII–VIII.

that is, casting three sixes at a first trial, and casting them again at the next trial, is

$$(1/216)^2 = (1/6)^6 = 1/46,656,$$

a very small probability indeed. Yet that man from the Basilicata kept on repeating the same extraordinary thing five times. Let us list the corresponding probabilities:

Repetitions	*Probability*
1	$1/6^3 = 1/216$
2	$1/6^6 = 1/46,656$
3	$1/6^9 = 1/10,077,696$
4	$1/6^{12} = 1/2,176,782,336$
5	$1/6^{15} = 1/470,184,984,576.$

Perhaps, the reverend adopted his initial assumption out of mere politeness; looking at the man from the Basilicata, he may have had his doubts about the fairness of the dice. The reverend remained silent after the three sixes turned up twice in succession, an event that under the initial assumption should happen not much more frequently than once in fifty thousand trials. He remained silent even longer. Yet, as the events became more and more improbable, attained and perhaps surpassed that degree of improbability that people regard as miraculous, the reverend lost patience, drew his conclusion, rejected his initial polite assumption, and spoke out forcibly.

(2) The anecdote that we have just discussed is interesting in just one aspect: it is typical. It shows clearly the circumstances under which we can reasonably reject a statistical hypothesis. We draw consequences from the proposed statistical hypothesis. Of special interest are consequences concerned with some event that appears *very improbable* from the standpoint of our statistical hypothesis; I mean an event the probability of which, computed on the basis of the statistical hypothesis, is very small. Now, we appeal to experience: we observe a trial that can produce that allegedly improbable event. If the event, in spite of its computed low probability, actually happens, it yields a strong *argument against* the proposed statistical hypothesis. In fact, we find it hard to believe that anything so extremely improbable could happen. Yet, undeniably, the thing did happen. Then we realize that any probability is computed on the basis of some statistical hypothesis and start doubting the basis for the computation of that small probability. And so there arises the argument against the underlying statistical hypothesis.

(3) As the reverend Galiani, we also felt obliged to reject the hypothesis of fair dice when we examined the extensive observations related in sect. 5 (1);

our reasons to reject it, however, were not quite as sharp as his. Could we find better reasons in the light of the foregoing discussion?

Here are the facts: 315,672 attempts to cast five or six spots with a dice produced 106,602 successes; see sect. 5 (2). *If* all dice cast were fair, the probability of a success would be 1/3. Therefore, we should expect about

$$315{,}672/3 = 105{,}224$$

successes in 315,672 trials. Thus, the observed number deviates from the expected number

$$106{,}602 - 105{,}224 = 1{,}378$$

units. Does such a deviation speak for or against the hypothesis of fair dice? Should we regard the deviation 1,378 as small or large? Is the probability of such a deviation high or low?

The last question seems to be the sensible question. Yet we still need a sensible interpretation of the short, but important, word "such." We shall reject the statistical hypothesis if the probability that we are about to compute turns out to be low. Yet the probability that the deviation should be exactly equal to 1,378 units is very small anyhow—even the probability of a deviation exactly equal to 0 would be very small. Therefore, we have to take into account *all the deviations of the same absolute value as, or of larger absolute value than, the observed deviation* 1,378. And so our judgment depends on the solution of the following problem: *Given that the probability of a success is* 1/3 *and that the trials are independent, find the probability that in* 315,672 *trials the number of successes should be either more than* 106,601 *or less than* 103,847.

With a little knowledge of the calculus of probability we find that the required probability is approximately

$$0.0000001983;$$

this means less than two chances in ten million. That is, an event has occurred that looks extremely improbable, *if* the statistical hypothesis is accepted that underlies the computation of probability. We find it hard to believe that such an improbable event actually occurred, and so the underlying hypothesis of fair dice appears extremely unlikely. Already in sect. 5 (1) we saw a good reason to reject the hypothesis of fair dice, but now we see a still better, more distinct, reason to reject it.

(4) The actual occurrence of an event to which a certain statistical hypothesis attributes a small probability is an argument against that hypothesis, and the smaller the probability, the stronger is the argument.

In order to visualize this essential point, let us consider the sequence

$$\frac{1}{10}, \quad \frac{1}{100}, \quad \frac{1}{1{,}000}, \quad \frac{1}{10{,}000}, \quad \ldots.$$

A statistical hypothesis implies that the probability of a certain event is 1/10.

The event happens. Should we reject the hypothesis? Under usual circumstances, most of us would not feel entitled to reject it; the argument against the hypothesis does not appear yet strong enough. If another event happens to which the statistical hypothesis attributes the probability 1/100, the urge to reject the hypothesis becomes stronger. If the alleged probability is 1/1,000, yet the event happens nevertheless the case against the hypothesis is still stronger. If the statistical hypothesis attributes the probability

$$\frac{1}{1,000,000,000}$$

to the event, or one chance in a billion, yet the event happens nevertheless, almost everybody would regard the hypothesis as hopelessly discredited, although there is no logical necessity to reject the hypothesis just at this point. If, however, the sequence proceeds without interruption so that events happen one after the other to which the statistical hypothesis attributes probabilities steadily decreasing to 0, for each reasonable person arrives sooner or later the critical moment in which he feels justified in rejecting the hypothesis, rendered untenable by its increasingly improbable consequences. And just this point is neatly suggested by the story of the reverend Galiani. The probability of the first throw of three sixes was 1/216; of the sequence of five throws of three sixes each, 1/470,184,984,576.

The foregoing discussion is of special importance for us if we adopt the standpoint that the theory of probability is a part of the study of nature. Any natural science must recur to observations. Therefore it must adopt rules that specify somehow the circumstances under which its statements are confirmed or confuted by experience. We have done just this for the theory of probability. We described certain circumstances under which we can reasonably consider a statistical hypothesis as practically refuted by the observations. On the other hand, if a statistical hypothesis survives several opportunities of refutation, we may consider it as corroborated to a certain extent.

(5) Probability, as defined in sect. 2, is the theoretical value of long range relative frequency. The foregoing gave us an opportunity to realize a few things. First, such a theoretical value depends, of course, on our theory, on our initial assumptions, on the statistical hypothesis adopted. Second, such a theoretical value may be very different from the actual value.

A suitable notation may help us to clarify our ideas. Let P be the probability of an event E computed on the basis of a certain statistical hypothesis H. Then P depends both on E and on H. (In fact, we could use, instead of P, the more explicit symbol $P(E, H)$ that emphasizes the dependence of P on E and H.)

In some of the foregoing applications we took the hypothesis H for granted (at least for the moment) and, computing P on the basis of H, we tried to predict the observable frequency of the event E. Yet, in the present section, we proceeded in another direction. Having observed the event E, we computed P on the basis of the statistical hypothesis H and, in view of the value of P obtained, we tried to judge the reliability of the hypothesis H. We perceive here a new aspect of P. The *smaller* P is, the more we feel inclined to reject the hypothesis H, and the *more unlikely* the hypothesis H appears to us: P indicates the likelihood of the hypothesis H. We shall say henceforward that P is the *likelihood of the statistical hypothesis* H, judged in view of the fact that the event E has been observed.

This terminology, which agrees essentially with the usage of statisticians, emphasizes a certain aspect of the dependence of P on the event E and the statistical hypothesis H. Our original terminology lays the stress on the complementary aspect of the same dependence: P is the *probability of the event* E, computed on the basis of the statistical hypothesis H.

Some practice in the use of this double terminology is needed to convince us that its advantages sufficiently outweigh its dangers.

8. Choosing between statistical hypotheses. The following example may provide a first orientation to the applications of the theory of probability in statistical research.

(1) A consumer buys a certain article from the producer in large lots. The consumer is a big consumer, a large merchandizing or industrial firm, or a government agency. The producer is also big and manufactures the article in question on a large scale. The article can be a nail, or a knob, or anything manufactured; an interesting example is a fuze, used for firing explosives in ammunition or in blasting operations. The article has to meet certain specifications. For example, the nail should not be longer than 2.04 inches nor shorter than 1.96 inches, its thickness is similarly specified, and perhaps also its minimum breaking strength; the burning time of the fuze is specified, and so on. An article that does not meet the specifications is considered as *defective*. Even the most carefully manufactured lot may contain a small fraction of defectives. Therefore the lot has to be inspected before it passes from the producer to the consumer. The lot may be fully inspected, that is, each article in the lot may be tested whether it meets the agreed specifications. Such a full inspection would be impractical for a lot of 10,000 nails and it would be preposterous for a lot of fuzes even if the lot is small; in order to measure its burning time, you have to destroy the fuze and there is not much point in destroying the whole lot by inspecting it. Therefore in many cases instead of inspecting the whole lot before acceptance, only a relatively small sample is taken from the lot. A simple procedure of such acceptance sampling is characterized by the following rule.

"Take a random sample of n articles from the submitted lot of N articles. Test each article in the sample. If the number of defectives in the sample does not exceed a certain agreed number c, the so-called *acceptance number*, the consumer accepts the lot, but he rejects it, and the producer takes it back, if there are more defectives than c in the sample."

The results obtained by this rule depend on chance. By chance, the fraction of defectives in the sample can be much lower or much higher than in the whole lot. If the sample turns out to be better than the lot, chance works against the consumer, and it works against the producer if the sample turns out to be worse than the lot. In spite of these risks, some such procedure appears necessary, and the rule formulated may be quite reasonable. We have to find out how the procedure works, how its result depends on the quality of the submitted lot. And so we are led to formulate the following problem: *Given p, the probability that an article chosen at random in the submitted lot is defective, find a, the probability that the lot will be accepted.*

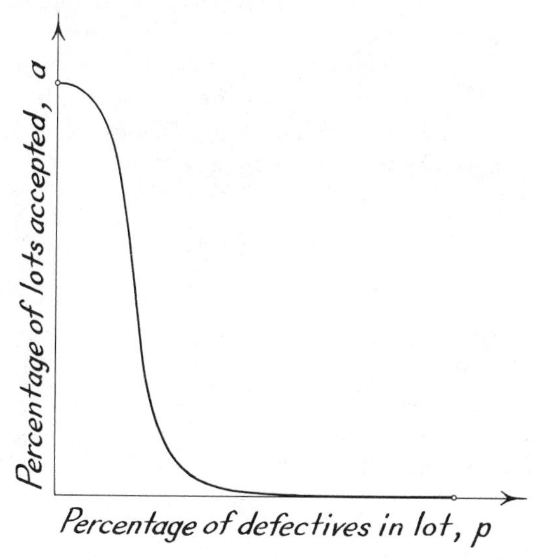

Fig. 14.4. Operating characteristic of an acceptance sampling procedure.

In the most important practical cases N, the size of the lot, is large even in comparison with n, the size of the sample. In such cases we may assume that N is infinite; we lose little in precision and gain much in simplicity. Assuming $N = \infty$, we easily find that

$$a = (1-p)^n + \binom{n}{1}p(1-p)^{n-1} + \binom{n}{2}p^2(1-p)^{n-2} + \dots + \binom{n}{c}p^c(1-p)^{n-c}.$$

We take this expression of a, the probability of acceptance, for granted and we concentrate on discussing some of its practical implications.

We graph a as function of p; see fig. 14.4. If we graphed $100a$ as function of $100p$, the form of the curve would be the same. Now, $100p$ is the percentage of defective articles in the lot submitted. On the other hand, if several lots with the same percentage of defectives were subjected to the same inspection procedure, the relative frequency of acceptance, that is, the ratio of accepted lots to submitted lots, would be close to a. Therefore, in the long run, $100a$ will be the percentage of the lots accepted among the lots submitted. This explains the labeling of the axes in fig. 14.4. The curve in fig. 14.4 allows us to survey how the procedure operates on lots of various quality, and so it is appropriately called the *operating characteristic.*

Judged by its effects, does the procedure appear reasonable? This is the question that we wish to consider.

If there are no defectives in the lot, there should be no chance for rejecting it. In fact, if $p = 0$ our formula yields $a = 1$, as it should. If there are only defectives in the lot, there should be no chance for accepting it. In fact, if $p = 1$ our formula yields $a = 0$, as it should. Both extreme points of the operating characteristic curve are obviously reasonable.

If the number of defectives increases, the chances of acceptance should diminish. In fact, differentiating with a little skill, we easily find the surprisingly simple expression

$$\frac{da}{dp} = -(n - c) \binom{n}{c} p^c (1 - p)^{n-1-c}$$

which is always negative. Therefore, the operating characteristic is necessarily a falling curve, as represented in fig. 14.4, which is again as it should be.

The absolute value of the derivative, or $-da/dp$, has also a certain practical significance. The change dp of the abscissa represents a change in the quality of the lot. The change da of the ordinate represents a change in the chances of acceptance, due to the change in quality. The larger is the ratio of these chances da/dp in absolute value, the sharper is the distinction made by the procedure between two slightly different lots. Especially, the point at which da/dp attains its maximum absolute value may be appropriately called the "point of sharpest discrimination." This point is easily recognized in the graph: it is the point of inflexion, if there is one, and otherwise the left-hand extremity of the curve. (Its abscissa is $p = c/(n - 1)$.)

(2) The rule appears sensible also from another standpoint. It has a certain flexibility. By choosing n, the size of the sample, and c, the acceptance number, we can adapt the rule to concrete requirements. Both the consumer and the producer require protection against the risks inherent in sampling. A bad lot may sometimes yield a good sample and a good lot a bad sample, and so there are two kinds of risks: the sampling procedure may accept a bad lot or reject a good lot. The consumer is against accepting bad lots and the producer is against rejecting good lots. Still, both kinds of undesirable decisions are bound to happen now and then and the only thing that we can reasonably demand is that they should not happen too often. This demand leads to concrete problems such as the following.

"Determine the sample size and the acceptance number so that there should be less than once chance in ten that a lot with 5% defectives is accepted and there should be less than five chances in a hundred that a lot with only 2% defectives is rejected."

In this problem, there are two unknowns, the sample size n and the acceptance number c. The condition of the problem requires the following two inequalities:

$$a > 0.95 \text{ when } p = 0.02,$$

$$a < 0.1 \quad \text{when } p = 0.05.$$

It is possible to satisfy these two simultaneous conditions, but it takes considerable numerical work to find the lowest sample size n and the corresponding acceptance number c for which the required inequalities hold.

We shall not discuss the numerical work. We are much more concerned here with visualizing the problem than with solving it. Let us therefore look a little further into its background. As we said already, both the acceptance of a bad lot and the rejection of a good lot are undesirable, the first from the consumer's viewpoint and the second from the producer's viewpoint. Yet the two undesirable possibilities may not be equally undesirable and the interests of consumer and producer may be not quite so sharply opposed. The acceptance of a bad lot is not quite in the interest of the producer; it may damage his reputation. Yet the rejection of a good lot may be very much against the interests of the consumer; he may need the articles urgently and the rejection may cause considerable delay. Moreover, repeated rejection of good lots, or even the danger of such rejection, may raise the price. If the interests of both parties are taken into account, the rejection of a good lot may be still less desirable than the acceptance of a bad lot. Seen against this background, it appears understandable that the conditions of our problem afford more protection against the rejection of the better quality than against the acceptance of the worse quality. (Only 5 chances in a hundred are allowed for the first undesirable event, but 10 chances in a hundred for the second.)

(3) The problem discussed under (2) admits another, somewhat different, interpretation.

The producer's lawyer affirms that there are no more than 2% defectives in the lot. Yet the consumer's lawyer contends that there are at least 5% defectives in the lot. For some reason (it may be a lot of fuzes) a full inspection is out of the question; therefore some sampling procedure has to decide between the two contentions. For this purpose the procedure outlined under (1) with the numerical data given in (2) can be appropriately used.

In fact, the conflicting contentions of the two lawyers suggest a fiction. We may pretend that there are exactly two possibilities with respect to the lot: the percentage of defectives in the lot is either exactly 2% or exactly

5%. Of course, nobody believes such a fiction, but the statistician may find it convenient: it restricts his task to a decision between two clear and simple alternatives. If the parties agree that the rejection of a lot with 2% defectives is less desirable than the acceptance of a lot with 5% defectives, the statistician may reasonably adopt the procedure outlined in (1) with the numerical data prescribed in (2). Whether the statistician's choice will satisfy the lawyers or the philosophers, I do not venture to say, but it certainly has a clear relation to the facts of the case. The statistician's rule, applied to a great number of analogous cases, accepts a good lot (with 2% defectives) about 950 times out of 1,000 and rejects it only about 50 times, but the rule rejects a bad lot (with 5% defectives) about 900 times out of 1,000 and accepts it only about 100 times. That is, the statistician's rule, which is based on sampling, cannot be expected to give the right decision each time, but it can reasonably be expected to give the right decision in an assignable percentage of cases *in the long run.*

(4) To give an adequate idea of what the statisticians are doing on the basis of just one example is, of course, a hopeless undertaking. Yet on the basis of the foregoing example we can obtain an idea of the statistician's task which, although very incomplete, is not very much distorted: the statistician designs rules of the same nature as the rule of acceptance sampling procedure outlined in (1) and considered in relation to numerical data in (2). We may understand the statistician's task if we have understood the nature of the rules he designs. Therefore, we have to formulate in general terms what seems to be essential in our particular rule; I mean the rule discussed in the foregoing sub-sections (1), (2), and (3).

Our rule prescribes a choice between two courses of action, acceptance and rejection. Yet the aspect of the problem considered under (3) is more suitable for generalization. There we considered a *choice between two statistical hypotheses.* (They were "this random sample is taken from a large lot with 2% defectives" and "this random sample is taken from a large lot with 5% defectives.") Any reasonable choice should be made with due regard to past experience and future consequences. In fact, our rule is designed with regard to both.

According to our rule, the choice depends upon a set of clearly specified observations (the testing of n articles and the number of defectives detected among the n articles tested). These observations constitute the relevant experience on which the choice is based. As our rule prefers a hypothesis to another on the basis of observations, it can claim to be named an *inductive* rule.

Our rule is designed with a view to probable consequences. The statistician cannot predict the consequences of any single application of the rule. He forecasts merely how the rule will work *in the long run.* If the choice prescribed by the rule is tried many times in such and such circumstances, it will lead to such and such result in such and such percentage of the trials, in the long run. Our rule is designed with a view to *long range consequences*.

To sum up, our *rule is designed to choose between statistical hypotheses, is based on a specified set of observations, and aims at long range consequences.* If we may regard our rule as sufficiently typical, we have an idea what the statisticians are doing: they are designing rules of this kind.

(In fact, they try to devise "best" rules of this kind. For example, they wish to render the chances of such and such undesirable effect a minimum, being given the size of the sample, on which the work and expense of the observations depend.)

(5) Taking a random sample from a lot is an important operation in statistical research. There is another problem about this operation that we have to discuss here. We keep our foregoing notation in stating the problem.

In a very large lot, $100p$ *percent of the articles is defective. In order to obtain some information about* p*, we take a sample of* n *articles from the lot, among which we find* m *defective articles. On the basis of this observation, which value should we reasonably attribute to* p*?*

There is an obvious answer, suggested by the definition of probability itself. Yet the problem is important and deserves to be examined from various angles.

Our observation yields some information about p. Especially, if m happens to be different from 0, we conclude that p is different from 0. Similarly, if m is less than n, we conclude that p is less than 1. Yet in any case p remains unknown and all values between 0 and 1 are eligible for p. If we attribute one of these values to p, we make a guess, we adopt a conjecture, we choose a statistical hypothesis.

Let us think of the consequences of our choice before we choose. If we have a value for p, we can compute the probability of the event the observation of which is our essential datum. I mean the probability for finding exactly m defective articles in a random sample of n articles. Let us call this probability P. Then

$$P = \binom{n}{m} p^m (1-p)^{n-m}.$$

The value of P depends on p, varies with p, can be greater or less. If, however, this probability P of an observed event is very small, we should reject the underlying statistical hypothesis. It would be silly to choose such an unlikely hypothesis that has to be rejected right away. Therefore let us choose the least unlikely hypothesis, the one for which the danger of rejection is least. That is, let us choose the value of p for which P is *as great as possible*.

Now, if P is a maximum, $\log P$ is also a maximum and, therefore,

$$\frac{d \log P}{dp} = \frac{m}{p} - \frac{n-m}{1-p} = 0.$$

This equation yields

$$p = \frac{m}{n}.$$

And so, after some consideration, we made the choice that we were tempted to make from the outset: as a reasonable approximation to p, the underlying probability, we choose m/n, the observed relative frequency.

Yet our consideration was not a mere detour. We can learn a lot from this consideration.

Let us begin by examining the rôle of P. This P is the probability of a certain observed event E (m defectives in a sample of size n). This probability is computed on the basis of the statistical hypothesis H_p that $100p$ is the percentage of defectives in the lot. The probability P varies with the hypothesis H_p (with the value of p). The smaller P, the less acceptable, the less likely appears H_p. Thus we are led to consider P as indicating the *likelihood* of the hypothesis H_p. This term "likelihood" has been introduced before (in sect. 7 (5)), in the same meaning, but now we may see the reasons for its introduction more clearly.

Let us emphasize that we choose among the various admissible statistical hypotheses H_p (with $0 \leqq p \leqq 1$) the one for which P, the likelihood of H_p, is as great as possible. Behind this choice there is a principle, appropriately called the *principle of maximum likelihood*, that guides the statistician also in other cases, less obvious than our case.

9. Judging non-statistical conjectures. We consider several examples in order to illustrate the same fundamental situation from several angles.

(1) The other day I made the acquaintance of a certain Mr. Morgenstern. This name is not very usual, but not unknown to me. There was a German author Morgenstern for whose nonsense poetry I have a great liking. And, Oh yes, my cousin who lives in Atlanta, Georgia, recently began work in the offices of Mark Morgenstern & Co., consulting engineers.

At the beginning I had no thoughts about Mr. Morgenstern. After a while, however, I hear that he is in the engineering business. Then other pieces of information leaked out. I hear that the first name of my new acquaintance is Mark, and that his place of business is Atlanta, Georgia. At this stage it is very difficult not to believe that this Mr. Morgenstern is the employer of my cousin. I ask Mr. Morgenstern directly and find that it is so.

This trivial little story is quite instructive. (It is based, by the way, on actual experience, but the names are changed, of course, and also some irrelevant circumstances.) That two different persons should have exactly the same last name is not improbable, provided that the name is very common such as Jones or Smith. It is more improbable that two different persons have the same first and last name, especially, when it is an uncommon name, such as Mark Morgenstern. That two different persons have the same profession, or the same large town as residence, is not improbable. Yet it is very improbable that two different persons taken at random should have the same unusual name, the same home town, and the same occupation. A chance coincidence was hard to believe and so my conjecture about my recent acquaintance Mr. Morgenstern was quite reasonable. It turned out to be correct, but this has really little to do with the merits of the case. My conjecture was reasonable, defensible, justifiable on the basis of the probabilities considered. Even if my conjecture had turned out incorrect, I would have no reason to be ashamed of it.

In this example, no numerical value was given for the probability decisively connected with the problem, but a rough estimate for it could be obtained with some trouble.

(2) Two friends who met unexpectedly decided to write a postcard to a third friend. Yet they were not quite sure about the address. Both remembered the city (it was Paris) and the street (it was Boulevard Raspail) but they were both uncertain about the number. "Wait," said one of the friends, "let us think about the number without talking, and each of us will write down the number when he thinks that he has got it." This proposal was accepted and it turned out that both remembered the same number: 79 Boulevard Raspail. They put this address on the postcard which eventually reached the third friend. The address was correct.

Yet what was the reason for adopting the number 79? By not talking to each other, the two friends made their memories work independently. They both knew that Boulevard Raspail is long enough to have buildings numbered at least up to 100. Therefore, it seems reasonable to assume that the probability for a chance coincidence of the two numbers is not superior to 1/100. Yet this probability is small, and so the hypothesis of a chance coincidence appears unlikely. Hence the confidence in the number 79.

(3) According to the statement of the bank, the balance of my checking account was $331.49 at the end of the past month. I compute my balance for the same date on the basis of my notes and find the same amount. After this agreement of the two computations I am satisfied that the amount in which they both agree is correct. Is this certain? By no means. Although both computations arrived at the same result, the result could be wrong and the agreement may be due to chance. Is that likely?

The amount, expressed in cents, is a number with five digits. If the last digit was chosen at random, it could just as well be 0 or 1 or 2, . . . or 8 as 9, and so the probability that the last digit should be 9 is just 1/10. The same is true for each of the other figures. In fact, if all figures were chosen at random, the number could be any one of the following:

000.00, 000.01, 000.02, . . . 999.99

I have here obviously 100,000 numbers. If that assemblage of five figures, 33149, was produced in some purely random way, all such assemblages could equally well arise. And, as there are 100,000 such assemblages, the probability that any one given in advance should be produced is

$$\frac{1}{100{,}000} = \left(\frac{1}{10}\right)^5 = 10^{-5}$$

Now, $10^{-5} = 0.00001$ is a very small probability. If, trying to produce an effect with such a small probability, somebody manages to succeed at the very first trial, the outcome may easily appear as miraculous. I am, however, not inclined to believe that there is anything miraculous about my modest bank account. A chance coincidence is hard to believe and so I am driven to the conclusion that the agreement of the two computations is due to the correctness of the result. Ordinary normal people generally think so in similar circumstances and after the foregoing considerations this kind of belief appears rather reasonable.

(4) To which language is English more closely related, to Hungarian or to Polish? Very little linguistic knowledge is enough to answer this question, but it is certainly more fun to obtain the answer by your own means than to accept it on the authority of some book. Here is a common sense access to the answer.

Both the form and the meaning of the words change in the course of history. We can understand the changes of form if we realize that the same language is differently pronounced in different regions, and we can understand the changes of meaning if we realize that the meaning of words is not rigidly fixed, but shifting, and changes with the context. In the second respect, however, there is one conspicuous exception: the meaning of the numerals one, two, three, . . . certainly cannot shift by imperceptible degrees. This is a good reason to suspect that the numerals do not change their meaning in the course of linguistic history. Let us, therefore, base a first comparison of the languages in question on the numerals alone. Table III lists the first ten numerals in English, Polish, Hungarian, and seven other modern European languages. Only languages that use the Roman alphabet are considered (this accounts for the absence of Russian and modern Greek). Certain diacritical marks (accents, cedillas) which are unknown in English are omitted (in Swedish, German, Polish, and Hungarian).

Looking at Table III and observing how the same numeral is spelled in different languages, we readily perceive various similarities and coincidences. The first five languages (English, Swedish, Danish, Dutch, and German) seem to be pretty similar to each other, and the next three languages (French, Spanish, and Italian) appear to be in even closer agreement; so we have two groups, one consisting of five languages, the other of three. Yet even these two groups appear to be somehow related; observe the coinciding spelling of 3 in Swedish, Danish and Italian, or that of 6 in English and French.

Table III. Numerals in ten languages.

English	Swedish	Danish	Dutch	German	French	Spanish	Italian	Polish	Hungarian
one	en	en	een	ein	un	uno	uno	jedem	egy
two	tva	to	twee	zwei	deux	dos	due	dwa	ketto
three	tre	tre	drie	drei	trois	tres	tre	trzy	harom
four	fyra	fire	vier	vier	quatre	cuatro	quattro	cztery	negy
five	fem	fem	vijf	funf	cinq	cinco	cinque	piec	ot
six	sex	seks	zes	sechs	six	seis	sei	szesc	hat
seven	sju	syv	zeven	sieben	sept	siete	sette	siedem	het
eight	atta	otte	acht	acht	huit	ocho	otto	osiem	nyolc
nine	nio	ni	negen	neun	neuf	nueve	nove	dziewiec	kilenc
ten	tio	ti	tien	zehn	dix	diez	dieci	dziesiec	tiz

Polish seems to be closer to one group in some respects, and to the other in other respects; compare 2 in Swedish and Polish, 7 in Spanish and Polish. Yet Hungarian shows no such coincidences with any of the nine other languages. These observations lead to the impression that Hungarian has little relation to the other nine languages which are all in some way related to each other. Especially, and this is the answer to our initial question, English seems to be definitely closer related to Polish than to Hungarian.

Yet there are several objections. A first objection is that "similarity" and "agreement" are vague words; we should say more precisely what we mean. This objection points in the right direction. Following its suggestion, we sacrifice a part of our evidence in order to render the remaining part more precise. We consider only the *initials* of the numerals listed in Table III. We compare two numerals expressing the same number in two different languages; we call them "concordant" if they have the same initial, and "discordant" if the initials are different. Table IV contains the number of concordant cases for each pair of languages. For instance, the

Table IV. Concordant initials of numerals in ten languages.

E	8	8	3	4	4	4	4	3	1	39
	Sw	9	5	6	4	4	4	3	2	45
		Da	4	5	4	5	5	4	2	46
			Du	5	1	1	1	0	2	22
				G	3	3	3	2	1	32
					F	8	9	5	0	38
						Sp	9	7	0	41
							I	6	0	41
								P	0	30
									H	8

number 7, in the same row as the letters "Sp" and in the same column as the letter "P" indicates that Spanish and Polish have exactly seven concordant numerals out of the possible 10 cases. The reader should check this and a few other entries of Table IV. The last column of Table IV shows how many concordant cases each language has with the other nine languages altogether. This last column shows pretty clearly the isolated position of Hungarian: it has only 8 concordant cases altogether whereas the number of concordant cases varies between 22 and 46 for the other nine languages.

Yet, perhaps, any definite conclusion from such data is rash: those coincidences of initials may be due to chance. This objection is easy to raise, but not so easy to answer. Chance could enter the picture through various channels. There may be an element of chance due to the fact that the correspondence between letters and pronunciation is by no means rigid. This is true even of a single language (especially of English). *A fortiori*, the same letter is often pretty differently pronounced in different languages and, on the other hand, different letters are sometimes very similarly pronounced. We have to admit that the coincidences observed are not free from some random element. Yet the question is: Is it *probable* that such coincidences as we have observed are due to mere chance?

If we wish to answer this question precisely, numerically, we have to adopt some precise, numerically definite statistical hypothesis and draw consequences from it which can be confronted with the observations. Yet the choice of a suitable hypothesis is not too obvious. We consider here two different statistical hypotheses.

I. There are two bags. Each bag contains 26 balls, each ball is marked with a letter of the alphabet, and different balls in the same bag are differently marked. With both hands, I draw simultaneously from both bags, one ball from each. The two letters so drawn may coincide or not; their coincidence is likened to the coincidence of the initials of the same numeral written in two different languages (and non-coincidence is likened to non-coincidence). The probability of a coincidence is 1/26.

II. The coincidence of the initials of the same numeral written in two different languages is again likened to the coincidence of two letters drawn simultaneously from two different bags and, again, both bags are filled in the same way with balls marked with letters. Yet now each of the bags contains 100 balls and each letter of the alphabet is used to mark as many different balls in the bag as there are numerals in Table III having that letter as initial. The probability of a coincidence is found to be 0.0948.

On both hypotheses, the comparison of the ten first numerals is likened to ten independent drawings of the same nature.

We can compare both hypotheses with the observations if we compute suitable probabilities. Tables V and VI contain the relevant material.

Table V compares the relative frequencies actually found with the probabilities computed. Columns (2) and (3) of Table V refer to all 45 pairs of languages considered in Table IV. Columns (4) and (5) of Table V refer only to 9 pairs, formed by Hungarian matched with the remaining nine languages. For the sake of concreteness, let us focus on the line that deals with 6 or more coincidences ($n = 6$). Such coincidences turned up in

Table V. Absolute and relative frequencies, and probabilities, for *n* or more coincidences of initials

(1)	(2)	(3)	(4)	(5)	(6)	(7)
	Frequencies				Probabilities	
n	10 languages		9 lang. v. Hu.		Hyp. II	Hyp. I
0	45	1.000	9	1.000	1.000000	1.000000
1	40	0.889	5	0.556	0.630644	0.324436
2	35	0.778	3	0.333	0.243824	0.054210
3	31	0.689	0	0.000	0.061524	0.005569
4	25	0.556	0	0.000	0.010612	0.000381
5	15	0.333	0	0.000	0.001281	0.000018
6	9	0.200	0	0.000	0.000108	0.000001
7	7	0.156	0	0.000	0.000006	0.000000

9 out of 45 cases as column (2) shows. Therefore, the observed relative frequency of 6 or more coincidences is $9/45 = 0.2$, whereas this many coincidences have only a little more than one chance in ten thousand to happen on hypothesis II, and only one chance in a million on hypothesis I; see columns (6) and (7), respectively. Similar remarks apply to the other lines of Table V: what has been actually observed appears as extremely improbable on either hypothesis, so there are strong grounds to reject both hypotheses. Yet columns (4) and (5) present a different picture: the coincidences observed are somewhat improbable on hypothesis I, but they appear as quite usual and normal from the standpoint of hypothesis II. The

Table VI. Total number of coincidences, observed and theoretical (Hypothesis II).

	Coincidences		Deviations	
	Observed	Expected	Actual	Standard
10 languages	171	42.66	128.34	7.60
9 lang. v. Hu.	8	8.53	– 0.53	2.78

impression gained from Table V is corroborated by Table VI: if we consider all 45 pairs of languages, the actually observed total number of coincidences exceeds tremendously what we have to expect on the basis of hypothesis II, yet the expected and observed numbers agree closely if we consider only the 9 pairs in which Hungarian is matched with the other 9 languages. (On hypothesis I, we have considerably stronger disagreement in both cases.)

In short there is no obvious interpretation of "chance" that would permit us to make chance responsible for all the coincidences observable in Table III; there are too many of them. Yet we can quite reasonably make chance responsible for the coincidences between Hungarian and the other languages. The explanation that Hungarian is unrelated to the other languages which are all related to each other has been vindicated.

The point is that this explanation has been vindicated, thanks to the consideration of probabilities, by so *few observations*. The explanation itself is supported by an overwhelming array of philological evidence.

(5) From appropriate observations (with telescope and spectroscope) we can conclude that certain elements found in the crust of our globe are also present in the sun and in certain stars. This conclusion is based on a physical law discovered by G. Kirchhoff almost a century ago (which says roughly that a luminous vapor absorbs precisely the same kind of light that it emits). Yet the conclusion appeals also to probabilities, and this is the point with which we are concerned here; we shall reduce the physical part of the argument to a schematic outline.

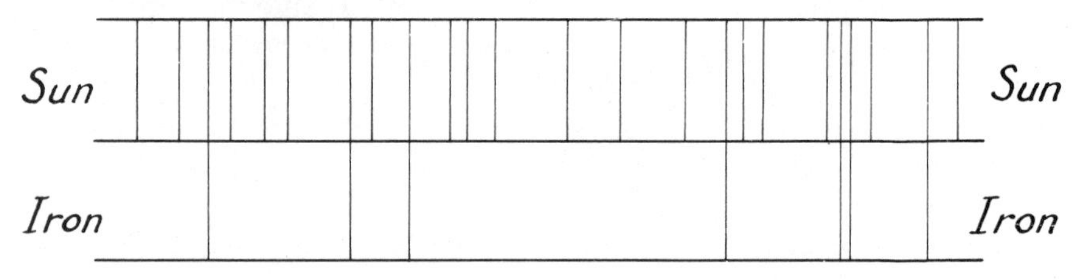

Fig. 14.5. Coincidences.

Using suitable apparatus (a prism or a diffraction grating) we can detect a sequence of lines in the light of the sun (in the solar spectrum). We can detect a sequence of lines also in the light emitted by certain substances, such as iron, vaporized at high temperature in the laboratory. (In fact, the lines in the spectrum of the sun, the Fraunhofer lines, are dark, and the lines in the spectrum of iron are bright.) Kirchhoff examined 60 iron lines and found that each of these lines coincides with some solar line. (See the rough schematic fig. 14.5 or *Encyclopaedia Britannica*, 14th edition, vol. 21, fig. 3 on plate I facing p. 560.) These coincidences are fully understandable if we assume that there is iron in the sun. (More exactly, these coincidences follow from Kirchhoff's law on emission and absorption if we assume that in the atmosphere of the sun there is iron vapor that absorbs some of the light emitted by the central part of the sun glowing at some still higher temperature.) Yet, perhaps (here is again that ever-present objection) these coincidences are due to chance.

The objection deserves serious consideration. In fact, no physical observation is absolutely precise. Two lines which we regard as coincident could be different in reality and just by chance so close to each other that, with the limited precision of our observations, we might fail to recognize their difference. We have to concede that any observed coincidence may be only an apparent coincidence and there may be, in fact, a small difference. Yet let us ask a question: Is it *probable* that each of the 60 coincidences observed springs from a random difference so small that it failed to be detected by the means of observation employed?

Kirchhoff, who registered the observed lines on an (arbitrary) centimeter scale, estimated that he could not have failed to recognize a difference that exceeded 1/2 millimeter on his scale. On this scale the average distance between two adjacent lines of the solar spectrum was about 2 millimeters. If the 60 lines of iron were thrown into this picture at random, independently from each other, what would be the probability that each falls closer to some solar line than 1/2 millimeter?

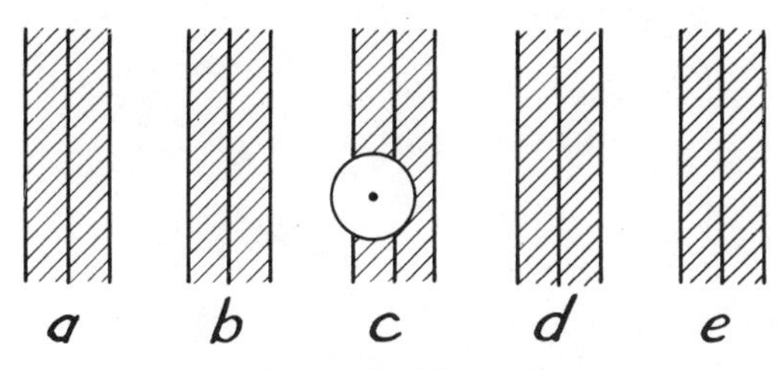

Fig. 14.6. Equidistant lines.

We bring this question nearer to its solution by formulating an equivalent question in a more familiar domain. Parallel lines are drawn on the floor; the average distance between two adjacent lines is 2 inches. We throw a coin on the floor 60 times. If the diameter of the coin is 1 inch, what is the probability that the coin covers a line each time?

In this last formulation, the question is easy to answer. Assume first that the lines on the floor are equidistant (as in fig. 14.6) so that the distance from each line to the next is 2 inches. If the coin covers a line, the center of the coin is at most at 1/2 inch distance from the line and, therefore, this center lies somewhere in a strip 1 inch wide that is bisected by the line (shaded in

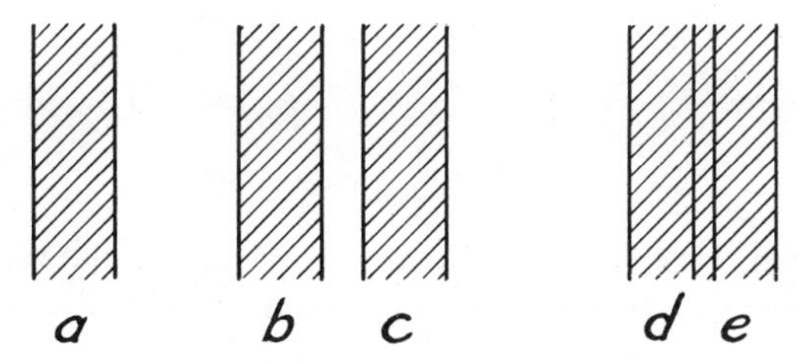

Fig. 14.7. Lines at irregular distances.

fig. 14.6). Obviously, the probability that the coin cast on the floor should cover a line is 1/2. The probability that the coin, cast on the floor 60 times, should cover some line each time, is $(1/2)^{60}$.

Assume now that the lines on the floor are not equidistant; the average distance between two adjacent lines is still supposed to be 2 inches. We imagine that the lines, which were equidistant originally, came into their present position by being shifted successively. If a line (as line *b* in fig. 14.7) is shifted so that its distance from its next neighbor remains more than 1 inch, the chances of the coin for covering some line remain unchanged. If, however, the line is so shifted (as line *d* in fig. 14.7) that its distance from the next line becomes less than 1 inch, the two (shaded) attached strips overlap and the chances of the coin to cover a line are diminished. Therefore, the required probability is less than $(1/2)^{60}$.

To sum up, if the iron lines were thrown by blind chance into the solar spectrum, the probability of the 60 coincidences observed by Kirchhoff would be less than 2^{-60} and so less than 10^{-18} or

$$\frac{1}{1,000,000,000,000,000,000}.$$

"This probability" says Kirchhoff, whom we quoted already in the motto prefixed to this chapter "is rendered still smaller by the fact that the brighter a given iron line is seen to be, the darker, as a rule, does the corresponding solar line appear. Hence this coincidence must be produced by some cause, and a cause can be assigned which affords a perfect explanation of the observed facts."

(6) The following example is not based on actual observation, but it illustrates a frequently arising, typically important situation.

An extremely dangerous disease has been treated in the same locality by two different methods which we shall distinguish as the "old treatment" and the "new treatment." Of the 9 patients who have been given the old treatment 6 died and only 3 survived, whereas of the 11 patients who received the new treatment only 2 died and 9 survived. The twenty cases are clearly displayed in Table VII.

Table VII. Four Place Correlation Table.

Patients	Died	Survived	Total
Old treatment	6	3	9
New treatment	2	9	11
Total	8	12	20

A first glance at this table may give us the impression that the observations listed speak strongly in favor of the new treatment. The relative frequency of fatal cases is

6/9 or 67% with the old treatment,

2/11 or 18% with the new treatment.

On second thoughts, however, we may wonder whether the observed numbers are large enough to give us any reasonable degree of confidence in the percentages just computed, 67% and 18%. Still, the fact remains that the number of fatal cases was much lower with the new treatment. Such a low mortality, however, could be due to chance. *How easily can chance produce such a result?*

This last question seems to be the right question. Yet, at any rate, the question must be put more precisely before it can be answered. We have to explain the precise meaning in which we used the words "chance" and "such." The word "chance" will be explained if we assimilate the present case to some suitable game of chance. A fair interpretation of the words "such a result" seems to be the following: we consider all outcomes in which the number of fatalities with the second treatment is *not higher than that actually observed.* Thus, we may be led eventually to the following formulation.

There are two players, Mr. Oldman and Mr. Newman, and 20 *cards, of which* 8 *are black and* 12 *are red. The cards are dealt so that Mr. Oldman receives* 9 *cards and Mr. Newman receives* 11 *cards. What is the probability that Mr. Newman receives* 2 *or less black cards?*

This formulation expresses as simply and as sharply as possible the contention that we have to examine: the difference between the old and the new treatment does not really matter, does not really influence the mortality, and the observed outcome is due to mere chance.

The required probability turns out to be

$$\frac{335}{8398} = 0.0399 \sim \frac{1}{25}$$

That is, an outcome that appears to be as favorable to the new treatment, as the observed outcome, or even more favorable, will be produced by chance about once in 25 trials. And so the numerical evidence for the superiority of the new treatment above the old cannot be simply dismissed, but is certainly not very strong.

In order to see clearly in these matters, let us give a moment's consideration to a situation in which the numerical data would lead us to a probability 1/10,000 instead of 1/25. Such data would make very hard to believe that the observed difference in mortality is due to mere chance but, of course, they would not prove right away the superiority of the new treatment. The data would furnish a pretty strong argument for the existence of *some non-random difference* between the two kinds of cases. What the nature of this difference actually is, the numbers cannot say. If only young or vigorous people received the new treatment, and only elderly or weak people the old treatment, the argument in favor of the medical superiority of one treatment above the other would be extremely weak.

(7) I think that the reader has noticed a certain parallelism between the six preceding examples of this section. Now this parallelism may be ripe to be brought into the open and formulated in general terms. Yet let us follow as far as possible the example of the naturalist who carefully compares the relevant details, rather than the example of those philosophers who rely mainly on words. We went into considerable detail in discussing our examples; if we do not take into account the relevant particulars carefully, our labor is lost.

In each example there is a *coincidence* and an *explanation*. (Name, surname, occupation and home town of a person I met coincide with those of a person I heard of. Explanation: the two persons are identical.—Two numbers, remembered or computed by two different persons, coincide. Explanation: the number, arrived at by two persons working independently, is correct.—The initials of several couples of numerals, designating the same number in two different languages, coincide. Explanation: the two languages are related.—The bright lines in the spectrum of iron, observed in laboratory experiments, coincide with certain dark lines in the spectrum of the sun. Explanation: there is iron vapor in the atmosphere of the sun.—A new treatment of a disease coincides with lower mortality. Explanation: the new treatment is more effective.)

Contrasting with these specific explanations, the nature of which varies with the nature of the example, there is another explanation which can be stated in the same terms in all examples: the observed coincidences are due to chance.

The specific explanations are not groundless, some of them are reasonably convincing, but none of them is logically necessary or rigidly proven. Therefore the situation is fundamentally the same in each example: there are two rival conjectures, a specific conjecture, and the "universally applicable" hypothesis of "randomness" which attributes the coincidences to chance.

Yet, if we look at it more closely, we perceive that the "hypothesis of randomness" is vague. The statement "this effect is due to chance" is ambiguous, since chance can operate according to different schemes. If we wish to obtain some more definite indication from it, we have to make the hypothesis of randomness more precise, more specific, express it in terms of probability, in short, we have to raise it to the rank of a *statistical hypothesis*.

In everyday matters we usually do not take the trouble to state a statistical hypothesis with precision or to compute its likelihood numerically. Yet we may take a first step in this direction (as in example (1)) or go even a little further (as in examples (2) or (3)). In scientific questions, however, we should clearly formulate the statistical hypothesis involved and follow it up to a numerical estimate of its likelihood, as in examples (5) and (6).

In the transition from the general and therefore somewhat diffuse idea of randomness to a specific statistical hypothesis we have to make a choice. There are cases in which we scarcely notice this choice, since we can perceive just one statistical hypothesis that is simple enough and fits the case reasonably well; in such a case the hypothesis chosen appears "natural" (as in examples (3), (5) and (6)). In other cases the choice is quite noticeable; we do not see immediately a statistical hypothesis that would be simple enough and fit the case somewhat "realistically," so we choose after more or less hesitation (as in example (4)).

Eventually there are two rival conjectures facing each other: a non-statistical, let us say "physical," conjecture *Ph* and a statistical hypothesis *St*. Now, a certain event *E* has been observed. This event *E* is related both to *Ph* and to *St*, and is so related that its happening could influence our choice between the two rival conjectures *Ph* and *St*. If the physical conjecture *Ph* is true, *E* appears as easily explicable, its happening is easily understandable. In the clearest cases (as in example (5)) *E* is implied by *Ph*, is a consequence of *Ph*. On the other hand, from the standpoint of the statistical hypothesis *St*, the event *E* appears as a "coincidence" the probability p of which can be computed on the basis of the hypothesis *St*. If the probability p of *E* turns out to be low, the happening of the event *E* is not easily explicable by "chance," that is by the statistical hypothesis *St*; this weakens our confidence in *St* and, accordingly, strengthens our confidence in *Ph*. On the other hand, if the probability p of the observed event *E* is high, *E* may appear as explicable by chance, that is, by the statistical hypothesis *St*; this strengthens somewhat our confidence in *St* and accordingly weakens our confidence in *Ph*.

It should be noticed that the foregoing is in agreement with what we said about rival conjectures in sect. 13.12 and adds some precision to the pattern of plausible reasoning discussed in sect. 12.3.

The omnipresent hypothesis of randomness is an alternative to any other kind of explanation. This seems to be deeply rooted in human nature. "Was it intention or accident?" "Is there an assignable cause or merely chance coincidence?" Some question of this kind occurs in almost every debate or deliberation, in trivial gossip and in the law courts, in everyday matters and in science.

10. Judging mathematical conjectures. We compare some examples treated in foregoing chapters with each other and with those treated in the foregoing section.

(1) Let us remember the story of a remarkable discovery told in sect. 2.6. Euler investigated the infinite series

$$1+\frac{1}{4}+\frac{1}{9}+\frac{1}{16}+\frac{1}{25}+\ldots+\frac{1}{n^2}+\ldots.$$

First he found various transformations of this series. Then, using one of these transformations, he obtained an approximate numerical value for the sum of the series, the value 1.644934. Finally, by a novel and daring procedure, he guessed that the sum of the series is $\pi^2/6$. Euler felt himself that his procedure was daring, even objectionable, yet he had a good reason to trust his discovery: the value found by numerical computation, 1.644934, coincided, as far as it went, with the value guessed

$$\frac{\pi^2}{6}=1.64493406\ldots.$$

And so Euler was confident. Yet was this confidence reasonable? Such a coincidence may be due to chance.

In fact, it is not outright impossible that such a coincidence is due to chance, yet there is just one chance in ten million for such a coincidence to happen: the probability that chance, interpreted in a simple manner, should produce such a coincidence of seven decimals is 10^{-7}; cf. sect. 9 (3) and ex. 11. And so we should not blame Euler that he rejected the explanation by chance coincidence and stuck to his guess $\pi^2/6$. He proved his guess ultimately. Yet we need not insist here on the fact that it has been proved: with or without confirmation, Euler's guess was, in itself, not only brilliant but also reasonable.

(2) Let us look again at sect. 3.1 and especially at fig. 3.1 which displays nine polyhedra. For each of these polyhedra we determined F, V, and E, that is, the number of faces, vertices, and edges, respectively, and listed the numbers found in a table (Vol. I, p. 36). Then we observed a regularity: throughout the table

$$F + V = E + 2.$$

It seemed to us improbable that such a persistent regularity should be mere coincidence, and so we were led to conjecture that the relation observed in nine cases is generally true.

There is a point in this reasoning that could be made more precise: what is the probability of such a coincidence? To answer this question, we have to propose a definite statistical hypothesis. I was not able to think of one that fits the case perfectly, but there is one that has some bearing on the situation. Let me state it in setting $F - 1 = X$, $V - 1 = Y$, $E = Z$. With this change of notation, the conjectural relation obtains the form $X + Y = Z$.

We have three bags, each of which contains n balls, numbered $1, 2, 3, \ldots n$. We draw one ball from each bag and let X, Y, and Z denote the number from the first, the second, and the third bag, respectively. What is the probability that we should find the relation

$$X + Y = Z$$

between the three numbers X, Y, and Z, produced by chance?

It is understood that the three drawings are mutually independent. With this proviso the probability required is determined and we easily find that it is equal to

$$\frac{n-1}{2n^2}.$$

Let us apply this to our example. Let us focus on the moment when we succeed in verifying the hypothetical relation for a new polyhedron. For example, after the nine polyhedra that we examined initially (in sect. 3.1) we took up the case of the icosahedron (in sect. 3.2). For the icosahedron, as we found, $F = 20$, $V = 12$, $E = 30$, and so, in fact

$$(F - 1) + (V - 1) = 19 + 11 = 30 = E.$$

Is this merely a random coincidence? We apply our formula, taking $n = 30$ (we certainly could not make n less than 30) and find that such an event has the probability

$$\frac{29}{2 \times 30^2} = \frac{29}{1800} = 0.016111;$$

that is, it has a little less prospect than 1 chance in 60. We may hesitate whether we should, or should not, ascribe the verification of the conjectured relation to mere chance. Yet if we succeed in verifying it for another polyhedron, with F, V, E about as large as for the icosahedron, and we are inclined to regard the two verifications as mutually independent, we face an event (the joint verification in both cases) with a probability less than $(1/60)^2$; this event has less chance to happen than 1 in 3600 and is, therefore, even harder to explain by chance. If the verifications continue without interruption, there comes a moment, sooner or later, when we feel obliged to reject the explanation by chance.

(3) In the foregoing example we should not stress too much the numerical values of the probabilities that we computed. To realize that the probability steadily decreases as verification follows verification may be more helpful in guiding our judgment than the numerical values computed. At any rate, there are cases in which it would be hard to offer a fitting statistical hypothesis and so it is not possible to compute the probabilities involved numerically, yet the calculus of probability still yields helpful suggestions.

In sect. 4.8 we compared two conjectures concerning the sum of four squares. Let us call them conjecture A and conjecture B, respectively. Conjecture A (that we have discovered at the end of sect. 4.6) asserts a remarkable rule that precisely determines in how many ways an integer of a certain form can be represented as a sum of four odd squares. Conjecture B (Bachet's conjecture) asserts that any integer can be represented as the sum of four squares in one or more ways. Each of the two conjectures offers a prediction about the sum of four squares, but the prediction offered by A is more precise than that offered by B. Just to stress this point, let us consider for a moment a quite unbelievable assumption. Let us assume that we know from some (mysterious) source that, in a certain case, the number of representations has an equal chance to have any one of the $r + 1$ values $0, 1, 2, \ldots r$, and cannot have a value exceeding r, which is a quite large number (and this should hold both under the circumstances specified in A and under those specified in B—a rather preposterous assumption). Now, A predicts that the number of representations has a definite value; B

predicts that it is greater than 0. Therefore, the probability that A turns out to be true in that assumed case is $1/(r+1)$, whereas the probability that B turns out to be true is $r/(r+1)$. In fact, both A and B turn out to be true in that case, both conjectures are verified, and the question arises which verification yields the stronger evidence. In view of what we have just discussed, it is much more difficult to attribute the verification of A to chance, than the verification of B. By virtue of this circumstance (in accordance with all similar examples discussed in this chapter) the verification of the more precise prediction A should carry more weight than the verification of the less precise prediction B. In sect. 4.8 we arrived at the same view without any explicit consideration of probabilities.

EXAMPLES AND COMMENTS ON CHAPTER XIV

First Part

Each example in this first part begins with a reference to some section or subsection of this chapter and supplies formulas or derivations omitted in the text. The solutions require some knowledge of the calculus of probability.

1. [Sect. 3 (3)] Accept the scheme of sect. 3 (3) for representing the succession of rainy and rainless days. Say "sunny" instead of "rainless," for the sake of convenience, and let r_r, s_r, r_s, and s_s denote probabilities,

r_r for a rainy day after a rainy day,
s_r for a sunny day after a rainy day,
r_s for a rainy day after a sunny day, and
s_s for a sunny day after a sunny day.

(a) Show that $r_r - r_s = s_s - s_r$.

(b) It was said that "a rainy day follows a rainy day more easily than a rainless day." What does this mean precisely?

2. [Sect. 3 (4)] It was said that "each letter tends to be unlike the foregoing letter." What does this mean precisely?

3. [Sect. 5 (1)] Find the general expression for the numbers in column (3) of Table I.

4. [Sect. 5 (2)] Find the general expression for the numbers in column (5) of Table I.

5. [Sect. 5 (3)] (a) Find the general expression for numbers in column (3) of Table II. (b) In order to detect a systematic deviation, if there is one, examine the differences of corresponding entries (on the same row) of columns (4) and (5); list the signs.

6. [Sect. 7 (1)] If a trial consists in casting three fair dice and a success consists in casting six spots with each dice, what is the probability of n successes in n trials?

7. [Sect. 7 (2)] Among the various events reported in the story of the Reverend Galiani told in sect. 7 (1), which one constitutes the strongest argument against the hypothesis of fair dice?

8. [Sect. 7 (3)] (a) Write down the formula that leads to the numerical value $1.983 \cdot 10^{-7}$.

(b) The probability of a success is 1/3. Find the probability that 315672 trials yield precisely 315672/3 successes.

9. [Sect. 8 (1)] The expression given for a is a sum. Each term of this sum is, in fact, a probability: for what?

10. [Sect. 8 (1)] Find the abscissa of the point of inflection of the curve represented by fig. 14.4.

11. [Sect. 9 (3)] Given a number of n figures. A sequence of n figures is produced at random, perhaps by a monkey playing with the keys of an adding machine. What is the probability that the sequence so produced should coincide with the given number? [Is the answer mathematically determined?]

12. [Sect. 9 (4)] Explain the computation of the probability 0.0948.

13. [Sect. 9 (4)] Find the general expression for the numbers (a) in column (6), (b) in column (7), of Table V.

14. [Sect. 9 (4)] Explain the computation of the expected numbers of coincidences in Table VI: (a) 42.66, (b) 8.53.

15. [Sect. 9 (4)] Explain the computation of the standard deviation 2.78 in the last row and last column of Table VI.

16. [Sect. 9 (5)] Why $(1/2)^{60}$?

17. [Sect. 9 (6)] Explain the computation of the probability 0.0399. [Generalize.]

18. [Sect. 10 (2)] Derive the expression $(n - 1)/2n^2$ for the required probability.

Second Part

19. *On the concept of probability.* Sect. 2 does not define what probability "is," it merely tries to explain what probability aims at describing: the "long range" relative frequency, the "final stable" relative frequency, or the relative frequency in a "very long" series of observations. How long such a series is supposed to be, was not stated. This is an omission.

Yet such omissions are not infrequent in the sciences. Take the oldest physical science, mechanics, and the definition of velocity in non-uniform, rectilinear motion: velocity is the space described by the moving point in a certain interval of time, divided by the length of that interval, provided that the interval is "very short." How short such an interval is supposed to be is not stated.

Practically, you take the interval of time measured as short, or the statistical series observed as long, as your means of observation allow you. Theoretically you may pass to the limit. The physicists, in defining velocity, let the interval of time tend to zero. R. von Mises, in defining probability, lets the length of the statistical series tend to infinity.

20. *How not to interpret the frequency concept of probability.* The D. Tel. shook his head as he finished examining the patient. (D. Tel. means doctor of teleopathy; although strenuously opposed by the medical profession, the practice of teleopathy has been legally recognized in the fifty-third state of the union.) "You have a very serious disease," said the D. Tel. "Of ten people who have got this disease only one survives." As the patient was sufficiently scared by this information, the D. Tel. went on. "But you are lucky. You will survive, because you came to me. I have already had nine patients who all died of it."

Perhaps the D. Tel. meant it. His grandfather was a sailor whose ship was hit by a shell in a naval engagement. The sailor stuck his head through the hole torn by the shell in the hull of the ship and felt protected "because," he reasoned, "it is very improbable that a shell will hit the same spot twice."

21. An official, charged to supervise an election in a certain locality, found 30 fake registrations among the 38 that he examined the first morning. A daily paper declares that at least 99% of the registrations in that locality are correct and above suspicion. How does the daily's assertion stand up in the light of the official's observation?

22. In the window of a watchmaker's shop there are four cuckoo clocks, all going. Three clocks out of the four are less than two minutes apart: can you rely on the time that they show? There is a natural conjecture: the clocks were originally set on time, but they are not very precise (they are just cuckoo clocks) and one is out of order. If this is so, you could rely on the time shown by three. Yet there is a rival conjecture, of course: those three clocks agree by mere chance. What is the probability of such an event?

23. If a, b, c, d, e, and f are integers chosen at random, not exceeding in absolute value a given positive integer n, what is the probability that the system

$$ax + by = e$$

$$cx + dy = f$$

of two equations with two unknowns has just one solution?

24. *Probability and the solution of problems.* In a crossword puzzle one unknown word with 5 letters is crossed by two unknown words with four letters each. You guess that the unknown 5 letter word is TOWER and then you have the situation indicated by the following diagram:

In order to test your guess, you would like to find one or the other four letter word crossing the conjectural TOWER. One of the crossing words could verify the O, the other the E. Which verification would carry more weight? And why?

25. *Regular and Irregular.* Compare the two columns of numbers:

I	II
1005	1004
1033	1038
1075	1072
1106	1106
1132	1139
1179	1173
1205	1206
1231	1239
1274	1271
1301	1303

One of these two columns is "regular," the other "irregular." The regular column contains ten successive mantissas from a four-place table of common logarithms. The numbers of the irregular column agree with the corresponding numbers of the regular column in the first three digits, yet the fourth digits could be the work of an unreliable computer: they have been chosen "at random." Which is which? [Point out an orderly procedure to distinguish the regular from the irregular.]

26. *The fundamental rules of the Calculus of Probability.* In computing probabilities we may visualize the set of possible cases and see intuitively that none is privileged among them, or we may proceed according to rules. It is important for the beginner to realize that he can arrive at the same result by these two different paths. The rules are particularly important when we regard the theory of probability as a purely mathematical theory.

The rules will be important in the next chapter. For all these reasons, let us introduce here the fundamental rules of the calculus of probability, using the bag and the balls;[8] cf. sect. 3.

The bag contains p balls. Some of the balls are marked with an A, others with a B, some with both letters, some are not marked at all. (There are p possible cases and two "properties," or "events," A and B.) Let us write $\bar{A}$ for the absence of A or "non-A." (We take — as the sign of negation, but place this sign on the top of the letter, not before it.) There are four possibilities, four categories of balls.

The ball has A, but has not B. We denote this category by $A\bar{B}$ and the number of such balls by a.

The ball has B, but has not A. We denote this category by $\bar{A}B$ and the number of such balls by b.

The ball has both A and B. We denote this category by AB and the number of such balls (common to A and B) by c.

The ball has neither A nor B. We denote this category by $\bar{A}\bar{B}$ and the number of such balls (different from those having A or B) by d.

Therefore, obviously,

$$a + b + c + d = p.$$

We let $\Pr\{A\}$ stand as abbreviation for the probability of A, and $\Pr\{B\}$ for that of B. With this notation, we have obviously

$$\Pr\{A\} = \frac{a + c}{p}, \qquad \Pr\{B\} = \frac{b + c}{p}.$$

Let $\Pr\{AB\}$ stand for the "probability of A and B," that is, the probability for the joint appearance of A and B. Obviously

$$\Pr\{AB\} = \frac{c}{p}.$$

Let $\Pr\{A \text{ or } B\}$ stand for the probability of obtaining A, or B, or both A and B.[9] Obviously

$$\Pr\{A \text{ or } B\} = \frac{a + b + c}{p}.$$

---

[8] We follow H. Poincaré, *Calcul des probabilités*, p. 35–39.

[9] The little word "or" has two meanings, which are not sufficiently distinguished by the English language, or by the other modern European languages. (They are, however, somewhat distinguished in Latin.) We may use "or" "exclusively" or "inclusively." "You may go to the beach or to the movies" (not to both) is *exclusive* "or" (in Latin "aut"). "You may go the beach or have a lot of candy" is *inclusive* "or" if you mean "one or the other or both." In legal or financial documents inclusive "or" is rendered as "and/or" (in Latin "vel"). In $\Pr\{A \text{ or } B\}$ we mean the *inclusive* "or."

We readily see that

$$\Pr\{A\} + \Pr\{B\} = \Pr\{AB\} + \Pr\{A \text{ or } B\}$$

and hence follows our first fundamental rule (the "or" rule):

$$\Pr\{A \text{ or } B\} = \Pr\{A\} + \Pr\{B\} - \Pr\{AB\}. \tag{1}$$

We wish now to define the *conditional* probability $\Pr\{A/B\}$, in words: probability of A *if* B (granted B, posito B, on the condition B, on the hypothesis B, . . .). Also this probability is intended to represent a long range relative frequency. We draw from the bag, repeatedly, one ball each time, replacing the ball drawn before drawing the next, as described at length in sect. 2 (1). Yet we *take into account only the balls having a B*. If among the first n such balls drawn, there are m balls that also have an A, m/n is the relative frequency that should be approximately, when n is sufficiently large, equal to $\Pr\{A/B\}$. It appears rather obvious that

$$\Pr\{A/B\} = \frac{c}{b+c}.$$

In fact, there are c balls having A among the $b + c$ balls having B; also the reasoning of sect. 2 (1) may be repeated; from a certain viewpoint, we could regard the expression of $\Pr\{A/B\}$ also as a definition. At any rate, we easily find, comparing the expressions of the probabilities involved, that

$$\Pr\{A/B\} = \Pr\{AB\}/\Pr\{B\}.$$

Interchanging A and B, we find the second fundamental rule (the "and" rule):

$$\Pr\{AB\} = \Pr\{A\}\Pr\{B/A\} = \Pr\{B\}\Pr\{A/B\}. \tag{2}$$

We can derive many other rules from (1) and (2). Observing that

$$\Pr\{A \text{ or } \bar{A}\} = 1, \qquad \Pr\{A\bar{A}\} = 0,$$

we obtain from (1), by substituting $\bar{A}$ for B, that

$$\Pr\{A\} + \Pr\{\bar{A}\} = 1, \tag{3}$$

what we could see also directly, of course. Similarly, since

$$\Pr\{AB \text{ or } \bar{A}B\} = \Pr\{B\}, \qquad \Pr\{(AB)(\bar{A}B)\} = 0,$$

we obtain from (1), by substituting AB for A and $\bar{A}B$ for B, that

$$\Pr\{B\} = \Pr\{AB\} + \Pr\{\bar{A}B\}. \tag{4}$$

We note here the following generalization of (2):

(5) $$\Pr\{AB/H\} = \Pr\{A/H\} \Pr\{B/HA\} = \Pr\{B/H\} \Pr\{A/HB\}.$$

We can also visualize (5) by using the bag and the balls.

27. *Independence.* We call two events independent of each other, if the happening (or not happening) of one has no influence on the chances of the other. Disregard, however, for the moment this informal definition and consider the two following formal definitions.

(I) A is called *independent of B* if

$$\Pr\{A/B\} = \Pr\{A/\bar{B}\}.$$

(II) A and B are called *mutually independent* if

$$\Pr\{A/B\} = \Pr\{A/\bar{B}\} = \Pr\{A\}, \qquad \Pr\{B/A\} = \Pr\{B/\bar{A}\} = \Pr\{B\}.$$

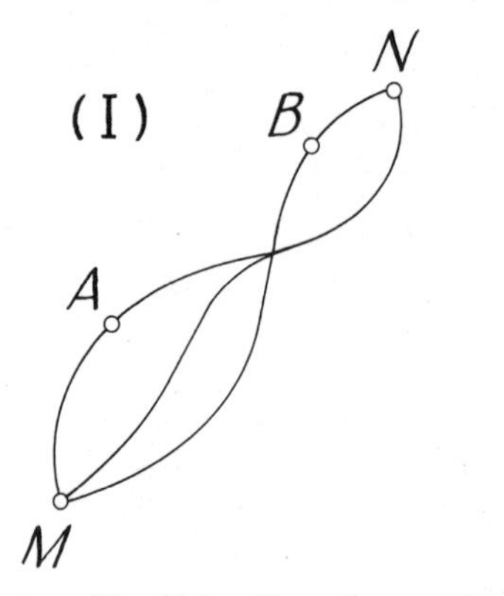

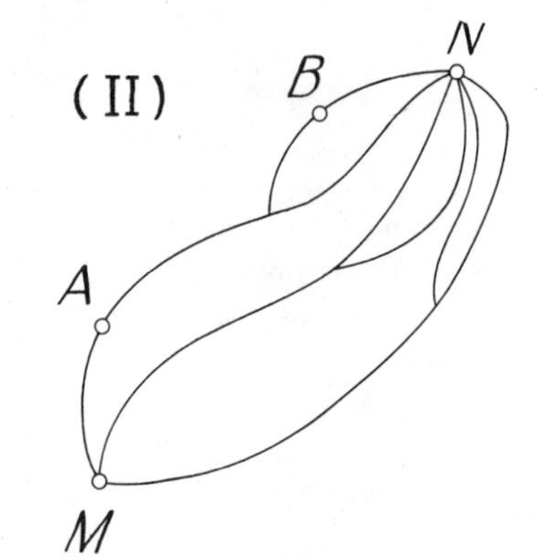

Fig. 14.8. Two systems of roads from the city M to the city N, with an essential difference.

Obviously, if A and B are mutually independent, A is independent of B. Using the rules of ex. 26, prove the theorem: *If none of the probabilities* $\Pr\{A\}$, $\Pr\{B\}$, $\Pr\{\bar{A}\}$, $\Pr\{\bar{B}\}$ *vanishes and any one of the two events A and B is independent of the other, they are mutually independent.*

28. Compare sect. 3 (5) with ex. 27.

29. A car traveling from the city M to the city N may pass through the town A and also through the town B. This is true of both systems of roads, (I) and (II), shown in fig. 14.8. Answer the following questions (a), (b), and (*c*) first in assuming that (I) represents the full system of roads between M and N, then in assuming the same thing about (II).

(a) Let A stand for the event that a car traveling from M to N passes through the town A, and B for the event that it passes through B. Assume (for both systems, (I) and (II)) that the three roads starting from M are equally well frequented (have the same probability) and also that the roads ending in N (there are 2 in (I), 6 in (II)) are equally well frequented. Find the probabilities $\Pr\{A\}$, $\Pr\{A/B\}$, $\Pr\{A/\bar{B}\}$, $\Pr\{B\}$, $\Pr\{B/A\}$, $\Pr\{B/\bar{A}\}$.

(b) Find $\Pr\{AB\}$ using the rule (2) of ex. 26.

(c) Verify that

$$\Pr\{A\} = \Pr\{B\}\Pr\{A/B\} + \Pr\{\bar{B}\}\Pr\{A/\bar{B}\},$$
$$\Pr\{B\} = \Pr\{A\}\Pr\{B/A\} + \Pr\{\bar{A}\}\Pr\{B/\bar{A}\}.$$

(d) What do you regard as the most important difference between (I) and (II)?

30. *Permutations from probability.* To decide the order in which the n participants should show their skill in an athletic contest, the name of each is written on a slip of paper, and then the n slips are drawn from a hat, one after the other, at random. What is the probability that the n names should appear in alphabetical order?

We present two solutions, and draw a conclusion from comparing them.

(1) Call E_1 the event that the slip drawn first is also alphabetically the first, E_2 the event that the slip drawn in the second place is also alphabetically the second, and so forth. The desired probability is

$$\begin{aligned}\Pr\{E_1 E_2 E_3 \ldots E_n\} &= \\ &= \Pr\{E_1\}\Pr\{E_2/E_1\}\Pr\{E_3/E_1E_2\}\ldots\Pr\{E_n/E_1\ldots E_{n-1}\} \\ &= \frac{1}{n}\cdot\frac{1}{n-1}\cdot\frac{1}{n-2}\cdots\frac{1}{1}\,.\end{aligned}$$

In fact, we obtain the first transformation by applying the rules (2) and (5) of ex. 26, and the second transformation by observing that there are n possible cases for E_1, $n - 1$ for E_2 after E_1, $n - 2$ for E_3 after E_1 and E_2, and so forth, whereas, for each of these events, there is just one favorable case.

(2) Call P_n the number of all the possible orderings (permutations, linear arrangements, . . .) of n distinct objects. The n names can come out from the hat in P_n ways, no one of these P_n possible cases appears as more privileged than the others, and among these P_n cases just one is favorable (the alphabetical order). Therefore, the desired probability is $1/P_n$.

(3) The results derived under (1) and (2) must be equal. Equating them, we evaluate P_n:

$$P_n = 1\cdot 2\cdot 3\ldots n = n!.$$

31. *Combinations from probability.* Mrs. Smith bought n eggs, not realizing that r of these eggs are rotten. She needs r eggs, and chooses as many among her n eggs at random. What is the probability that all r eggs chosen are rotten?

As in ex. 30 we present two solutions, and draw a conclusion from comparing them.

(1) Call E_1 the event that the first egg opened by Mrs. Smith is rotten, E_2 the event that the second egg is rotten, and so forth. The desired probability is

$$\begin{aligned}\Pr\{E_1 E_2 E_3 \ldots E_r\} &= \Pr\{E_1\} \Pr\{E_2/E_1\} \Pr\{E_3/E_1E_2\} \ldots \Pr\{E_r/E_1 \ldots E_{r-1}\} \\ &= \frac{r}{n} \cdot \frac{r-1}{n-1} \cdot \frac{r-2}{n-2} \cdots \frac{1}{n-r+1}.\end{aligned}$$

The first transformation is obtained by rules (2) and (5) of ex. 26, the second from the consideration of possible and favorable cases for E_1, for E_2 after E_1, and so on.

(2) We have a set of n distinct objects. Any r objects chosen among these n objects form a subset of size r of the given set of size n: call C_r^n the number of all such subsets. (Usually C_r^n is called the number of "combinations" of r things selected from among n things.) In the case of Mrs. Smith's eggs, there are C_r^n possible cases, no one more privileged than the others, and among these C_r^n cases just one is favorable (if getting rotten eggs is "favorable"). Hence the desired probability is $1/C_r^n$.

(3) Comparing (1) and (2), we evaluate C_r^n:

$$C_r^n = \frac{n(n-1)\ldots(n-r+1)}{1 \cdot 2 \quad \ldots \quad r} = \frac{n!}{r!(n-r)!} = \binom{n}{r}.$$

32. *The choice of a rival statistical conjecture: an example.* One person withdrew \$875 from his savings account on a certain date, and another person received \$875 two days later. The coincidence of these two amounts, one withdrawn, the other received, may be regarded as circumstantial evidence, as an indication that a crime has been committed; cf. ex. 13.6. If the jury finds it too hard to believe that this coincidence is due to mere chance, a conviction may result. Hence the problem: what is the probability of such a coincidence? The less the probability is, the more difficult it is to attribute the coincidence to chance, and the stronger is the case against the defendants.

Yet we cannot compute a probability numerically without assuming some definite statistical hypothesis. Which hypothesis should we assume? In a serious case we should give serious thought to such a question. Let us survey a few possibilities.

(1) As the number 875 has three digits, we may regard the positive integers with not more than three digits as admissible, and we may regard them as equally admissible. The probability that two such integers, chosen at random, independently from each other, should coincide, is obviously

1/999. This probability is pretty small—but is the assumption that underlies its computation reasonable?

(2) As 875 has less than five digits, we could regard all positive integers with less than five digits as equally admissible. This leads to the probability 1/9999 for the coincidence. This probability is very small indeed, but our assumption is far-fetched, even frivolous.

(3) If the point appears as important, the court can order inspection of the books of the bank or summon one of its competent officials to testify. And so it has been ascertained that immediately before the withdrawal of that sum $875 the amount $2581.48 was deposited on the account. In possession of this relevant information we may regard as possible and equally admissible cases the sums 1, 2, 3, . . . 2581 that could have been withdrawn from the account. Just one of these cases, 875, has to be termed favorable and so we are led to the probability 1/2581 for the coincidence. This is a small probability, but our assumption may seem reasonably realistic.

(4) We could have considered not only withdrawals in dollars, but also withdrawals in dollars and cents, such as $875.31. If we consider all such cases as equally admissible, the probability for a coincidence becomes 1/258148. This is a very small probability, but our assumption may appear less realistic: withdrawals in dollars and cents such as $875.31 are more usual from a checking account than from a savings account.

(5) On the contrary, one could argue that the amounts withdrawn from a savings account are usually "round" amounts, divisible by 100, or 50, or 25. Now, 875 is divisible by 25. If we regard only multiples of 25 as admissible, and equally admissible, the probability in question becomes 1/103.

Of course, we could imagine still other and more complicated ways to compute the probability, but we should not insist unduly on such a transparent example. The example served its purpose if the reader can see by now the following two points.

(a) Although some of the five assumptions discussed may seem more acceptable than others, no one is conspicuously superior to the others, and there is little hope to find an assumption that would be satisfactory in every respect and could be regarded as the best.

(b) Each of the five assumptions considered attributes a rather small probability to the coincidence actually observed, and so, on the whole, our consideration upholds the common sense view: "It is hard to believe that this coincidence is due to mere chance."

33. *The choice of a rival statistical conjecture: general remarks.* Let us try to learn something more general from the particular example considered (ex. 32). Let us reconsider the general situation discussed in sect. 14.9 (7). An event E has occurred and has been observed. Concerning this event, there are two rival conjectures facing each other: a "physical" conjecture P, and a statistical hypothesis H. If we accept the physical conjecture P, E is easily and not unreasonably explicable. If we accept the statistical hypothesis H, we can compute the probability p for the happening of such an event as E. If p is "small," we may be induced to reject the statistical hypothesis H. At any rate, the smallness of p weakens our confidence in H and therefore strengthens somewhat our confidence in the rival conjecture P.

Yet ex. 32 makes us aware that the quality of the statistical hypothesis H plays a role in the described reasoning. The statistical hypothesis H may appear as unnatural, inappropriate, far-fetched, frivolous, cheap, *unreliable* from the start. Or H may appear as natural, appropriate, realistic, reasonable, *reliable* in itself.

Now p, the probability of the event E computed on the basis of the hypothesis H, may be so small that we reject H: a rival of P drops out of the race. This increases the prospects of P—but it may increase them a lot or only a little: this depends on the quality of the rival. If the statistical hypothesis H appeared to us originally as appropriate and reliable, H was a dangerous rival and its fall strengthens P appreciably. If, however, H appeared to us as inappropriate and unreliable from the start, H was a weak rival; its fall is not surprising and strengthens P very little.

Being given a clear statistical hypothesis H, the probability p of the event E is determined, and the statistician can compute it. Yet the statistician's customer, who may be a biologist, or a psychologist, or a businessman, or any other non-statistician, has to decide what this numerical value of p means in his case. He has to decide how small a p is enough to reject or weaken the statistical hypothesis H. Yet the customer is usually not even directly interested in the statistical hypothesis H: he is primarily concerned with the rival "physical" conjecture P. And he has to decide how much weight the rejection or weakening of H has in strengthening P. This latter decision obviously cannot depend on the numerical value of p alone: it certainly depends on the choice of H.

I am afraid that the statistician's customer who wishes to make use of the numerical value p furnished by the statistician, without realizing the import of the statistical hypothesis H for his problem, just deceives himself. He can scarcely realize the import of H if he does not realize that his physical conjecture P could be also confronted with statistical hypotheses different from H. Cf. ex. 15.5.

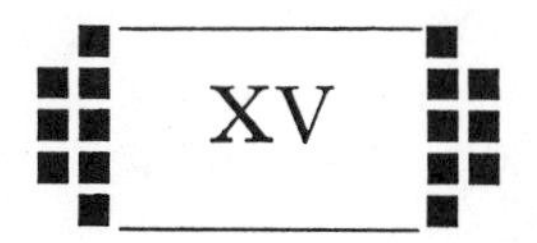

THE CALCULUS OF PROBABILITY AND THE LOGIC OF PLAUSIBLE REASONING

It is difficult to estimate the probability of the results of induction.—LAPLACE[1]

We know that the probability of a well-established induction is great, but, when we are asked to name its degree, we cannot. Common sense tells us that some inductive arguments are stronger than others, and that some are very strong. But how much stronger or how strong we cannot express.—JOHN MAYNARD KEYNES[2]

1. Rules of plausible reasoning? In the foregoing three chapters we collected patterns of plausible reasoning. Do these "patterns" constitute "rules" of plausible reasoning? How far and in which way are they binding, authoritative, imperative? There is a certain danger of losing ourselves in purely verbal explanations. Therefore I wish to consider the question more concretely, even a little personally.

(1) I remember a conversation on invention and plausible reasoning. It happened long ago. I talked with a friend who was much older than myself and could look back on a distinguished record of discoveries, inventions, and successful professional work. As he talked on plausible reasoning and invention, he doubtless knew what he was talking about. He maintained with unusual warmth and force of conviction that invention and plausible reasoning have no rules. Hunches and guesses, he said, depend on experience and intuition, but not on rules: there are no rules, there can be no rules, there should be no rules, and if there were some rules, they were useless anyway. I maintained the contrary—a conversation is uninteresting if there is no difference of opinion—yet I felt the strength of his position. My friend was a surgeon. A wrong decision of a surgeon may cost a life and

[1] Essai philosophique sur les probabilités; see *Oeuvres complètes de Laplace*, vol. 7, p. CXXXIX.

[2] *A treatise on probability*, p. 259.

sometimes, when a patient suddenly starts bleeding or suffocating, the right decision must come in a second. I understood that people who have to make such responsible quick decisions have no use for rules. The time is too short to apply a rule properly, and any set pattern could misguide you; what you need is intense concentration upon the situation before you. And so people come to distrust "rules" and to rely on their "intuition" or "experience" or "intuition-and-experience."

In the case of my friend, there was still something else, perhaps. He was a little on the domineering side. He hated to relinquish power. He felt, perhaps, that acknowledging a rule is like delegating a part of his authority to a machine, and so he was against it.

Let us note: distrust in rules of reasoning may come naturally to intelligent people.

(2) Two people presented with the same evidence may judge it very differently. Two jurors who sat through the same proceedings may disagree: one thinks that the evidence introduced is sufficient proof against the defendant and the other thinks that it is not. Such disagreement may have thousand different grounds: people may be moved in opposite directions by fears, hopes, prejudices and sympathies, or by personal differences. Perhaps, one of the jurors is stupid and the other is clever, or one slept through the proceedings and the other listened intently. Yet the personal differences underlying the disagreement may be more subtle. Perhaps both jurors are honest and reasonably unprejudiced, both followed the proceedings with attention, and both are intelligent, but in a different way. The first juror may be a better observer of demeanor. He observes the facial expressions of the witnesses, the tics of the defendant; he notices when an answer is haltingly given; he is impressed by quick motions of the eyes and little gestures of the hands. The other juror may be a less skillful observer of facial expressions, but a better judge of social relations: he understands better the milieu and the circumstances of the people involved in the case. Seeing the same things with different eyes, honestly and not unintelligently, the two jurors come to opposite conclusions.

Let us not neglect the obvious and let us note: two people presented with the same evidence may honestly disagree.

(3) My friend and I are both interested in the conjecture A. (This friend is a mathematician, and A is a mathematical conjecture.) We both know that A implies B. And now we find that B, this consequence of A, is true. We agree, as we honestly have to agree, that this verification of its consequence B is evidence in favor of the conjecture A, but we disagree about the value, or weight, of this evidence. One of us asserts that this verification adds very little to the credibility of A, and the other asserts that it adds a lot.

This disagreement would be understandable if we were very unequally familiar with the subject and one of us knew many more formerly verified consequences than the other. Yet this is not so. We know about the same consequences of A verified in the past. We agree even that there is little analogy between the just verified B and those formerly verified consequences. We agree also, as we honestly should, that this circumstance renders the evidence for A stronger. Yet one of us says "just a little stronger," the other says "a lot stronger," and we disagree.

We both suspected, even a short while ago, that B is false and it came to us as a surprise that B is true. In fact, from the standpoint of a rather natural assumption (or statistical hypothesis) B appears pretty improbable. We both perceive that this circumstance renders the evidence for A stronger. Yet one of us says "just a little stronger," the other says "a lot stronger," and we keep on disagreeing.

We are both perfectly honest, I think, and our disagreement is not merely a matter of temperament. We disagree because his *background* is different from mine. Although we had about the same scientific training, we developed in different directions. His work led him to distrust the hypothesis A. He hopes, perhaps, that one day he will be able to refute that conjecture A. As to myself, I do not dare to hope that I shall prove A one day. Yet I must confess that I would like to prove A. In fact, it is my ambition to prove A, but I do not wish to fool myself into the illusion that I shall ever be able to prove A. Such an incompletely avowed hope may influence my judgement, my evaluation of the weight of the evidence. Yet I may have other grounds besides: still more obscure, scarcely formulated, inarticulate grounds. And my friend may have some grounds too that he did not yet confess to himself. At any rate, such differences in our backgrounds may explain the situation: we disagree concerning the strength of the evidence, although we agree in all the clearly recognizable points that should influence the strength of the evidence in an impersonal judgement according to universally accepted reasonable standards.

Let us note: two persons presented with the same evidence and applying the same patterns of plausible inference may honestly disagree.

(4) We tried to see plausible reasoning at work, concretely, in the behavior of people facing concrete problems. We have now, I hope, a somewhat clearer idea in which way our patterns are "binding," how far they can be regarded as "rules."

Yet there are other approaches to explore. Formal Logic and the Calculus of Probability have clear strict rules which appear somehow related to our patterns. What is the nature of this relation? This is the question that we shall discuss in the next sections.

2. An aspect of demonstrative reasoning. A comparison of plausible reasoning with demonstrative reasoning may be useful at this stage. Yet of course the aspect of plausible reasoning at which we have arrived cannot stand comparison with the highly sophisticated stage at which the theory of demonstrative reasoning has arrived at this date, after a development of more than two thousand years of which the last fifty were particularly crowded. A comparison with a more primitive aspect of demonstrative reasoning may be more helpful. Let us put ourselves, more or less, in the position of a contemporary of Aristotle.

Aristotle noticed that reasoning conforms to certain *patterns*. He observed, I imagine, such patterns in philosophical or political or legal or everyday arguments, recognized the patterns as they occurred, extracted and formulated them. These patterns are the syllogisms. The examples by which Aristotle finds necessary to support his syllogisms seem to bear witness to the idea that he discovered his syllogisms by a sort of induction—and how could he have discovered them otherwise? At any rate, the idea that the syllogisms may have been discovered inductively brings them a little nearer to our patterns of plausible reasoning.

Instead of the "subsumptive" syllogism, so dear to Aristotle and still dearer to his scholastic followers, let us consider the "modus tollens" of the "hypothetical" syllogism, which we have already considered in sect. 12.1:

$$\frac{\begin{array}{c} A \text{ implies } B \\ B \text{ false} \end{array}}{A \text{ false}}$$

Even from a quite primitive standpoint, we can see various remarkable features in this pattern of reasoning: it is *impersonal, universal, self-sufficient*, and *definitive*.

(1) By using the word *impersonal* we emphasize that the validity of the reasoning does not depend on the personality of the reasoner, on his mood, or taste, or class, or creed, or color.

(2) By using the word *universal*, we emphasize that the statements considered (denoted by A and B) need not belong to this or that particular field of knowledge, to mathematics or physics, to law or logic, but can belong to any one of these fields, or to any field whatsoever. They can be concerned with any sufficiently clear object of human thought: the syllogistic conclusion applies to all such objects.

(3) In order to understand the next point, we should realize that our knowledge and our reasonable beliefs can be changed by new information. Yet there is something unchangeable in the syllogism considered. Having once accepted the premises we cannot avoid accepting the conclusion. At some later date, we may receive new information about the matters involved in our syllogistic reasoning. If, however, this information does not change our acceptance of the premises, it cannot reasonably change our acceptance of the conclusion. The inference of a demonstrative syllogism requires nothing from outside, is independent of anything not mentioned explicitly in the premises. In this sense, the syllogism is *self-sufficient*: nothing is needed beyond the premises to validate the conclusion and nothing can invalidate it if the premises remain solid.

This "self-sufficiency" or "autarky" of the syllogism is, perhaps, its most noteworthy feature. Let us quote Aristotle himself: "A syllogism is a discourse in which, certain things being stated, something other than what is stated follows of necessity from their being so. I mean by the last phrase that they produce the consequence, and by this, that no further term is required from without in order to make the consequence necessary."

(4) If the premises are unquestionably certain, we can "detach" the conclusion from the syllogism. That is, if you know for certain both that "A implies B" and that "B is false," you may forget about these premises and just keep the conclusion "A is false" as your *definitive* mental possession.

We have examined just one out of the several kinds of syllogisms, but the syllogism examined is typical: also the other syllogisms are impersonal, universal, self-sufficient, and definitive. And these features foreshadow the general character of demonstrative reasoning.

3. A corresponding aspect of plausible reasoning. Let us compare the pattern of demonstrative reasoning (the "modus tollens") discussed in the foregoing section with the pattern of plausible reasoning introduced in sect. 12.1:

$$\frac{\begin{array}{c} A \text{ implies } B \\ B \text{ true} \end{array}}{A \text{ more credible}}$$

Between these two patterns, the "demonstrative" and the "plausible," there is a certain outward similarity. (The demonstrative pattern is traditional, and the other has been fashioned after it, of course.) Yet let us compare them more thoroughly.

Both patterns have the same first premise

$$A \text{ implies } B.$$

The second premises

$$B \text{ false} \qquad\qquad B \text{ true}$$

are just opposite, but they are equally clear and definite; they are on the same logical level. Yet there is a great difference between the two conclusions

$$A \text{ false} \qquad\qquad A \text{ more credible}$$

These conclusions are on different logical levels. The conclusion of the demonstrative pattern is on the same level as the premises, but the conclusion of our pattern of plausible reasoning is of a different nature, less sharp, less fully expressed.

The plausible conclusion is comparable to a force which has *direction and magnitude*. This conclusion pushes us in a certain direction: A becomes *more* credible. This conclusion has also a certain strength: A may become *much more* credible or *just a little more* credible. The conclusion is not fully expressed and is not fully supported by the premises. *The direction is expressed and is implied by the premises, the strength is not.* For any reasonable person, the premises involve that A becomes more credible (certainly not less credible) but my friend and I may disagree *how much* more credible A becomes. *The direction is impersonal, the strength may be personal.* My friend and I may honestly disagree about the weight of the conclusion, since our temperaments, our backgrounds, and our unstated reasons may be different. Yet the strength of the conclusion matters. If two jurors judge differently the strength of a conclusion, one may be for acquittal and the other against it. If two scientists judge differently the strength of a conclusion, one may be for undertaking a certain experiment and the other against it.

The conclusion of our pattern of plausible inference appears as *one-sided* when compared with the actual beliefs and acts of the reasoning persons: it expresses merely one aspect, and neglects others. If we realize this, the nature of plausible reasoning may seem less baffling and elusive. At any rate we are now better prepared to compare the patterns, the demonstrative and the plausible, point by point. Each of the following subsections refers to the correspondingly numbered subsection of the foregoing sect. 2.

(1) When we are reasoning in accordance with our pattern of plausible inference, we conform to a principle: the verification of a consequence strengthens the conjecture. This principle seems to be generally recognized, independently of personal differences and idiosyncrasies. Thus our pattern appears as *impersonal.*

We pay, however, a price for such "impersonality." Our pattern succeeds in being impersonal because it is one-sided, restricted to one aspect of the plausible inference. When we raise the question "*How much* is the conjecture strengthened by the verification of this consequence?" we open the door to personal differences.

(2) We took pains to show by many examples in the foregoing chapters that we naturally follow our pattern of plausible inference in dealing with mathematical conjectures. The underlying principle is generally recognized in the natural sciences, and it is implicitly admitted in the law courts, and in everyday life. The verification of a consequence is regarded as reasonable evidence for a conjecture in any domain. Thus our pattern appears as *universal.*

We pay, however, a price for such "universality." Our pattern succeeds in being universal because it is one-sided, restricted to one aspect of plausible inference. The universality becomes blurred when we raise the question "What is the weight of such evidence?" In order to judge the weight of the evidence, you have to be familiar with the domain; in order to judge the weight with assurance, you have to be an expert in the domain. Yet you cannot be familiar with all domains, and you can still less be an expert in all domains. And so everyone of us will notice soon enough that there are practical limits to the universality of plausible inference.

(3) As far as it is expressed, the plausible conclusion is supported by the premises. On the basis of the evidence supplied by the premises it is reasonable to place more confidence in A. At some later date, however, we may receive new information which, without changing our reliance on the premises, may change our opinion about A: we may find A less credible, or we may even succeed in proving A false.

This does not constitute an objection against the pattern of reasoning: as far as the evidence expressed in the premises goes, the conclusion is justified. A verdict of the jury may condemn the innocent or acquit the criminal. Yet such injustice of the verdict may be justifiable: on the basis of the available evidence no better verdict was possible. Such is the nature of plausible inference, and so our pattern of plausible reasoning may be termed *self-sufficient.*

Yet this kind of self-sufficiency or "autarky" does not mean durability. Moreover, the weight of the evidence, which is not mentioned in the conclusion of our (one-sided) pattern, but is important nevertheless, depends on things which are not mentioned in the premises. The strength of the conclusion (not its direction) requires things outside the premises.

(4) We cannot "detach" the conclusion of our pattern of plausible reasoning. "A is rendered more credible" is meaningless without reference to the premises that explain by which circumstances it was rendered so. Referred to the premises, the plausible conclusion makes perfectly good sense and is perfectly reasonable, but it may diminish in value as time goes by, although the premises remain intact. The plausible conclusion may be very valuable in the moment when it emerges, but the advance of knowledge is likely to depreciate it: its importance is only momentary, transitory, ephemeral, *provisional.*

In short, our pattern of plausible reasoning is one-sided and leaves an ample margin for disagreement in things that matter. Yet, at the price of such one-sidedness, it manages to be impersonal and universal, even self-sufficient in a way. Still, it cannot escape being merely provisional.

It would be foolish to deplore that in several respects our pattern of plausible reasoning falls short of the perfection of demonstrative reasoning. On the contrary, we should feel a little satisfaction that we succeeded in clarifying somewhat a difference that we may have suspected from the beginning.

From the outset it was clear that the two kinds of reasoning have different tasks. From the outset they appeared very different: demonstrative reasoning as definite, final, "machinelike"; and plausible reasoning as vague, provisional, specifically "human." Now we may see the difference a little more distinctly. In opposition to demonstrative inference, plausible inference leaves indeterminate a highly relevant point: the "strength" or the "weight" of the conclusion. This weight may depend not only on clarified grounds such as those expressed in the premises, but also on unclarified unexpressed grounds somewhere in the background of the person who draws the conclusion. A person has a background, a machine has not. Indeed, you can build a machine to draw demonstrative conclusions for you, but I think you can never build a machine that will draw plausible inferences.

4. An aspect of the calculus of probability. Difficulties. A highly important step in the construction of a physical theory is its *formulation in mathematical terms*. We come to a point in our investigation where we should undertake such a step; we should formulate our views on plausible reasoning in mathematical terms.

No attempt at formulating a theory of plausible reasoning can disregard a historical fact: the calculus of probability was considered by Laplace and by many other eminent scientists as the appropriate expression of the rules of plausible inference. There are some grounds for this opinion and some objections against it. We begin by considering some of the difficulties.

We wish to use the calculus of probability to render more precise our views on plausible reasoning. Yet we could have some misgivings about such a procedure, for we have seen in the foregoing chapter that the calculus of probability is a (quite acceptable) theory of random mass phenomena. How could the calculus of probability be both the theory of mass phenomena and the logic of plausible inference?

This is not a strong objection; there is no real difficulty. The calculus of probability *could* be both things, *could* have two interpretations. In fact, a mathematical theory may have several different interpretations. The same differential equation (Laplace's equation) describes the steady irrotational flow of an incompressible non-viscous fluid and the distribution of forces in an electrostatic field. The same equation describes also the steady flow of heat, the steady flow of electricity, the diffusion of a salt dissolved in water under appropriate conditions, and still other phenomena. And so it is not excluded *a priori* that the same mathematical theory may serve two purposes. Perhaps, we may use the calculus of probability both in describing random mass phenomena and in systematizing our rules of plausible inference.

It is important, however, to distinguish clearly between these two interpretations. Thus, we may use the symbol $\Pr\{A\}$ (see ex. 14.26) in both interpretations, but only with certain safeguards, and we must understand quite clearly both meanings of the symbol, and see the difference between the two meanings.

In the foregoing chapter on random mass phenomena, we considered some kind of event A such as the birth of a boy, or the fall of a raindrop in some specified location, or the casting of a specified number of spots with a die, and so on. We used the symbol $\Pr\{A\}$ to denote the probability of the event A, that is, the theoretical value of the long range relative frequency of the event A.

In the present chapter, however, we have to deal with plausible reasoning. We consider some conjecture A, and we are concerned with the reliability of this conjecture A, the strength of the evidence in favor of A, our confidence in A, the degree of credence we should give to A, in short the *credibility of the conjecture A*. We shall take the symbol $\Pr\{A\}$ to denote the credibility of A.

Thus, in the present chapter we shall use the symbol $\Pr\{A\}$ in its second meaning as "credibility" unless we explicitly state the contrary. Such use of the symbol is not objectionable, but we have to discuss carefully the concept of credibility if we do not wish to expose ourselves to grave objections.

First, there is an ambiguity to avoid. The symbol $\Pr\{A\}$ should represent the credibility of A, or the strength of the evidence for the conjecture A. Such evidence is strong if it is convincing. It is convincing if it convinces somebody. Yet we did not say whom it should convince: you, or me, or Mr. Smith, or Mrs. Jones, or whom? The strength of the evidence could also be conceived *impersonally*. If we conceive it so, the degree of belief that you or me or any other person may happen to have in a proposed conjecture is irrelevant, but what matters is the *degree of reasonable belief* that anyone of us *should* have. We did not say yet, and we have still to decide, in what exact sense we should use the term "credibility of A" and the corresponding symbol $\Pr\{A\}$.

There is another difficulty. The magnitudes considered by the physicists such as "mass," "electric charge," or "reaction velocity" have an *operational* definition; the physicist knows exactly which operations he has to perform if he wishes to ascertain the magnitude of an electric charge, for example. The definition of "long range relative frequency," although in some way less distinct than that of an electric charge, is still operational; it suggests definite operations that we can undertake to obtain an approximate numerical value of such a frequency. The trouble with the concept of the "credibility of a conjecture" is that we do not know any operational definition for it. What is the credibility of the conjecture that Mr. Jones is unfaithful? This credibility may have at this moment a definite value in the mind of Mrs. Jones (a negligibly small value, we hope) but we do not know how to determine that value numerically. What is the credibility of the law of universal gravitation judged on the basis of the observations reported in the first edition of Newton's *Principia*? This question could be of high interest to some of us. (Not to Mrs. Jones, perhaps, but to Laplace or Keynes if they were still alive—see the quotations prefixed to this chapter.) But nobody dared to propose a definite numerical value for such a credibility.

We have still to give a suitable interpretation of the term "credibility of the conjecture A," and to the corresponding symbol $\Pr\{A\}$. This interpretation must be such that the difficulty of an operational definition does not interfere with it. Moreover, and this is the main thing, this interpretation should enable us to view the rules of plausible reasoning systematically and realistically.

5. An aspect of the calculus of probability. An attempt. You have just been introduced to Mr. Anybody and you have to say a few words to him. You two are really complete strangers to each other and so your conversation may be cautious. Still you cannot help touching upon various assertions such as "It will rain tomorrow," "The next Big Game will be won by The Blues," "Corporation So-and-so will pay a higher dividend next year," "Mrs. Somebody whose divorce is the talk of the town was unfaithful," "Polio is caused by a virus," or any other assertions, A, B, C, D, E, ... Mr. Anybody attaches to the assertion A a definite degree of credence $\Pr\{A\}$. If you are very clever, you can feel after some time spent with Mr. Anybody whether $\Pr\{A\}$ is low or high. Yet however clever you are, I cannot believe that you are able to ascribe a definite numerical value to $\Pr\{A\}$, the credibility of the statement A in the eyes of Mr. Anybody. (Although it would be interesting: the values of $\Pr\{A\}$, $\Pr\{B\}$, $\Pr\{C\}$, . . . could sharply characterize Mr. Anybody's personality.)

Let us be realistic and acknowledge the impossibility of a task that is obviously beyond our means: let us regard $\Pr\{A\}$, the credibility of the conjecture A in the eyes of Mr. Anybody, as a definite positive fraction

$$0 < \Pr\{A\} < 1$$

the numerical value of which, however, we *do not know*. And let us treat similarly $\Pr\{B\}$, $\Pr\{C\}$, . . . if B, C, . . . are conjectures, that is, clearly formulated (perhaps mathematical) statements of which, however, Mr. Anybody does not know at this time whether they are true or false. If, however, A is true and Mr. Anybody knows it, we set $\Pr\{A\} = 1$. If A is false and Mr. Anybody knows it, we set $\Pr\{A\} = 0$.

It seems to me that our ignorance of the numerical values of $\Pr\{A\}$, $\Pr\{B\}$, . . . cannot really hurt us. In fact, we are not concerned here with the personal opinions of Mr. Anybody. We are concerned with impersonal and universal rules of plausible inference. We wish to know, in the first place, whether there are such rules at all, and then we wish to know whether or not the calculus of probability does disclose such rules (as Laplace and others maintained). For the moment we rather hope that there are such rules, and that Mr. Anybody, as a sensible person, reasons according to such rules whatever degrees of credence $\Pr\{A\}$, $\Pr\{B\}$, $\Pr\{C\}$, . . . he may attach at this moment to the statements A, B, C, . . . discussed. And so I cannot see why our ignorance of the numerical values of $\Pr\{A\}$, $\Pr\{B\}$, $\Pr\{C\}$, . . . should hurt us.

Let us attempt, therefore, to apply the rules of the calculus of probability to the credibilities $\Pr\{A\}$, $\Pr\{B\}$, $\Pr\{C\}$, . . . as interpreted: positive fractions measuring degrees of confidence of that mythical or idealized person, Mr. Anybody. We wish to see whether, in doing so, we can elicit anything that can be reasonably interpreted as an impersonal and universal rule of plausible reasoning. Our attempt may fail, of course, but I cannot see at this moment why it should fail, and so I am cautiously hopeful.[3]

6. Examining a consequence. Mr. Anybody is investigating a certain conjecture A. This conjecture A is clearly formulated, but Mr. Anybody does not know whether A is true or not and wants badly to find out which is the case: is A true or is it false? He notices a certain consequence B of A. He is satisfied that

$$A \text{ implies } B.$$

Yet he does not know whether B is true or false, and sometimes, when he is tired of investigating A, he thinks of switching to the investigation of B.

We have considered this situation many times, and now we wish to reconsider it in the light of the calculus of probability. We wish to pay due attention to three credibilities: $\Pr\{A\}$, $\Pr\{B\}$, and $\Pr\{A/B\}$. Mr. Anybody knows very well that A and B are unproved and unrefuted, but he believes in them to a certain degree and this degree is expressed by $\Pr\{A\}$ and $\Pr\{B\}$, respectively. Also $\Pr\{A/B\}$, the degree of credence that he could place in A if he knew that B is true, plays an important role in his deliberations.

We have no means of assigning a numerical value to any of these credibilities, although we can sometimes imagine in which direction a change in the state of knowledge of Mr. Anybody would change the value of one or the other. At any rate, the calculus of probability yields a relation between them.

In fact, by one of the fundamental theorems on probability (see ex. 14.26 (2))

$$\Pr\{A\} \Pr\{B/A\} = \Pr\{B\} \Pr\{A/B\}.$$

Yet now, since A implies B, B must be true if A is true, and so

$$\Pr\{B/A\} = 1.$$

Hence we obtain

$$\Pr\{A\} = \Pr\{B\} \Pr\{A/B\}. \tag{I}$$

Let us visualize the contents of this equation.

(1) Mr. Anybody decided to investigate the consequence B of his conjecture A. He did not succeed yet in bringing this investigation to a conclusion. Yet sometimes he saw indications that B may be true, and sometimes

[3] That the Calculus of Probability should deal primarily with degrees of belief (confidence, confirmation, certitude, . . .) and not with more or less idealized relative frequencies, is the opinion of many authors among which I quote only two: J. M. Keynes, *A treatise on probability* (cf. especially p. 34, 66, 160), and B. de Finetti, La prévision, ses lois logiques, ses sources subjectives, *Annales de l'Institut Henri Poincaré*, vol. 7 (1937) p. 1–68. There is no space to explain my differences from these authors, or the differences between them, but I wish to express my thanks to both. The standpoint here adopted is similar to, but not quite the same as, that of my previous paper: Heuristic reasoning and the theory of probability, *American Math. Monthly*, vol. 48 (1941) p. 450–465.

indications to the contrary. His confidence in B, which we call $\Pr\{B\}$, rose and fell accordingly. Yet he did not observe anything that would have changed his views about the relation between A and B or about $\Pr\{A/B\}$. How did all this influence $\Pr\{A\}$, his confidence in A?

Equation (I) shows that $\Pr\{A\}$ *changes in the same direction as* $\Pr\{B\}$, provided that $\Pr\{A/B\}$ remains unchanged. This agrees with our former remarks, especially with those in sect. 13.6. (Observe, that we consider only the *direction* of the change that we can sometimes ascertain, and not its *amount* that we can never know precisely.)

(2) Mr. Anybody succeeded in proving B, which is a consequence of the conjecture A that he originally investigated. Before proving B, he had some confidence in B, the degree of which we have represented by $\Pr\{B\}$; he had also some confidence in A, of degree $\Pr\{A\}$. He sometimes considered $\Pr\{A/B\}$, the confidence he could place in A after a proof of B. After the proof of B, his confidence in B attains the maximum value 1, and his confidence in A becomes, of course, $\Pr\{A/B\}$. (Substituting 1 for $\Pr\{B\}$ in equation (I), we are led formally to the new value of the credibility of A.) We suppose here that his views about the relation between A and B and his evaluation of $\Pr\{A/B\}$ remain unchanged.

Observing that $0 < \Pr\{B\} < 1$, we derive from equation (I) the inequality

$$\text{(II)} \qquad \Pr\{A\} < \Pr\{A/B\}.$$

Now, $\Pr\{A\}$ and $\Pr\{A/B\}$ represent the credibility of A before and after the proof of B, respectively. Therefore inequality (II) is the formal expression of a principle with which we met so often: *the verification of a consequence renders a conjecture more credible*; cf. sect. 12.1, for instance.

(3) Yet we can learn still more from equation (I) which we write in the form

$$\text{(III)} \qquad \Pr\{A/B\} = \frac{\Pr\{A\}}{\Pr\{B\}}.$$

The left-hand side, the credibility of A after the verification of B, is expressed in terms of the confidence that the researcher had in A and B, respectively, before such verification. Let us compare various cases of successful verification of a consequence. These cases have one circumstance in common: the same confidence $\Pr\{A\}$ was placed in the conjecture A (at which the investigation aims) before the verification of its consequence B. Yet these cases differ in another respect: the consequence B (which was eventually verified) was expected with more confidence in some cases and with less confidence in others. That is, we regard $\Pr\{A\}$ as constant and $\Pr\{B\}$ as variable. How does the variation of $\Pr\{B\}$ influence the weight of the evidence resulting from the verification of the consequence B?

Let us pay due attention to the extreme cases. Since B is a consequence of A, B is certainly true when A is true, and so $\Pr\{B\}$, the credibility of B, cannot be less than $\Pr\{A\}$, the credibility of A. On the other hand no credibility can exceed certainty: $\Pr\{B\}$ cannot be greater than 1. We have determined the bounds between which $\Pr\{B\}$ is contained:

$$\Pr\{A\} \leqq \Pr\{B\} < 1.$$

The lower bound is attained, when not only A implies B, but also B implies A, so that the two assertions A and B are equivalent, stand and fall together, in which case they are, of course, equally credible. The upper bound 1 cannot be really attained: if it were attained, B would be certain before investigation, and we have not included this case in our consideration. Yet the upper bound can be approached: B can be almost certain before examination. How does the evidence resulting from the verification of B change when $\Pr\{B\}$ varies between its extreme bounds?

The evidence is stronger, when $\Pr\{A/B\}$, the new confidence in A resulting from the verification of the consequence B, is greater. It is visible from the relation (III) that

as $\Pr\{B\}$ decreases from 1 to $\Pr\{A\}$

$\Pr\{A/B\}$ increases from $\Pr\{A\}$ to 1.

This statement expresses in a new language a point that we have recognized before (sect. 12.3): *the increase of our confidence in a conjecture due to the verification of one of its consequences varies inversely as the credibility of the consequence before such verification.* The more unexpected a consequence is, the more weight its verification carries. The verification of the most surprising consequence is the most convincing, whereas the verification of a consequence that we did not doubt much anyway ($\Pr\{B\}$ almost 1) has little value as evidence.

(4) The situation just discussed can be viewed from another standpoint. Using some simple rules of the calculus of probability (cf. ex. 14.26 formulas (4), (2), (3), in this order) we obtain

$$\begin{aligned}\Pr\{B\} &= \Pr\{AB\} + \Pr\{\bar{A}B\} \\ &= \Pr\{A\}\Pr\{B/A\} + \Pr\{\bar{A}\}\Pr\{B/\bar{A}\} \\ &= \Pr\{A\} + [1 - \Pr\{A\}]\Pr\{B/\bar{A}\}.\end{aligned}$$

In passing to the last line, we also used that $\Pr\{B/A\} = 1$, which expresses that B is a consequence of A. In substituting for $\Pr\{B\}$ the value just derived, we obtain from (III)

$$\text{(IV)} \qquad \Pr\{A/B\} = \frac{\Pr\{A\}}{\Pr\{A\} + [1 - \Pr\{A\}]\Pr\{B/\bar{A}\}}.$$

Let us assume as before that $\Pr\{A\}$ is constant; that is, let us survey various cases in which the confidence in A, the conjecture examined, was the same before testing the consequence B of A. Yet $\Pr\{A/B\}$, the credibility of A after the verification of B, still depends on $\Pr\{B/\bar{A}\}$, the credibility of B (before verification, of course) viewed under the assumption that A is not true. And $\Pr\{B/\bar{A}\}$ can vary; it can take, in fact, any value between 0 and 1. Now, by our formula (III),

as $\Pr\{B/\bar{A}\}$ decreases from 1 to 0

$\Pr\{A/B\}$ increases from $\Pr\{A\}$ to 1.

This statement expresses in a new language a point that we have discussed before (sect. 13.10). Let us look at the extreme cases. If B without A is hardly credible ($\Pr\{B/\bar{A}\}$ almost 0) the verification of the consequence B brings the conjecture A close to certainty. On the other hand, the verification of a consequence B that we would scarcely doubt even if A were false ($\Pr\{B/\bar{A}\}$ almost 1) adds little to our confidence in A.

7. Examining a possible ground. After the broad and cautious discussion in the foregoing section we can proceed a little faster in surveying similar situations.

Here is such a situation: the aim of our research is a certain conjecture A. We notice a possible ground for A, that is, a proposition B from which A would follow:

A is implied by B.

We start investigating B. If we succeeded in proving B, A would also be proved. Yet B turns out to be false. How does the disproof of B affect our confidence in A?

Let the calculus of probability answer this question. Since A is implied by B

$$\Pr\{A/B\} = 1.$$

Let us combine this with some basic formulas (see ex. 14.26 (4), (2), (3)):

$$\begin{aligned}\Pr\{A\} &= \Pr\{AB\} + \Pr\{A\bar{B}\} \\ &= \Pr\{B\}\Pr\{A/B\} + \Pr\{\bar{B}\}\Pr\{A/\bar{B}\} \\ &= \Pr\{B\} + (1 - \Pr\{B\})\Pr\{A/\bar{B}\}.\end{aligned}$$

We obtain hence that

$$\text{(I)} \qquad \Pr\{A/\bar{B}\} = \frac{\Pr\{A\} - \Pr\{B\}}{1 - \Pr\{B\}}.$$

The left-hand side represents the credibility of A after B (which is a possible ground for A) has been refuted. The right-hand side refers to the situation before the refutation of B. By the way, this right-hand side can be transformed so that equation (I) appears in the form

$$\Pr\{A/\bar{B}\} = \Pr\{A\} - \Pr\{B\}\frac{1 - \Pr\{A\}}{1 - \Pr\{B\}}$$

and hence we see that

$$\text{(II)} \qquad \Pr\{A/\bar{B}\} < \Pr\{A\}.$$

Both sides of this inequality represent the credibility of the conjecture A, the left-hand side after the refutation of B, the right-hand side before this refutation. Therefore, inequality (II) expresses a rule: *our confidence in a conjecture can only diminish when a possible ground for the conjecture has been exploded.* (Cf. sect. 13.2.)

Yet we can learn more from equation (I). Let us consider $\Pr\{A\}$ as constant and $\Pr\{B\}$ as variable. That is, let us survey various cases which differ in one important respect: in the degree of our confidence in B. Our confidence in B may be very small, but it cannot be arbitrarily large: it can never exceed our confidence in A, since if B is true A is also true. (Yet A could be still true even if B is false.) And so, we determined the extreme values between which $\Pr\{B\}$ can vary:

$$0 < \Pr\{B\} \leqq \Pr\{A\}.$$

We see from equation (I) that

as $\Pr\{B\}$ increases from 0 to $\Pr\{A\}$

$\Pr\{A/\bar{B}\}$ diminishes from $\Pr\{A\}$ to 0.

That is, *the more confidence we placed in a possible ground of our conjecture, the greater will be the loss of faith in our conjecture when that possible ground is refuted.*

8. Examining a conflicting conjecture. We consider now another situation: we examine two conflicting conjectures, A and B. When we say that A conflicts with B or

A is incompatible with B

we mean that the truth of one of them implies the falsity of the other. We are, in fact, primarily concerned with A and we have started investigating B because we thought that the investigation of B could shed some light on A. In fact, a proof of B would disprove A. Yet we succeeded in disproving B. How does this result affect our confidence in A?

Let the calculus of probability give the answer. Let us begin by expressing in the language of this calculus that A and B are incompatible. This means in other words that A and B cannot both be true, and so

$$\Pr\{AB\} = 0.$$

Now we conclude from our basic formulas (cf. ex. 14.26 (4), (2), (3))

$$\begin{aligned}\Pr\{A\} &= \Pr\{AB\} + \Pr\{A\bar{B}\}\\ &= \Pr\{A\bar{B}\}\\ &= \Pr\{\bar{B}\}\Pr\{A/\bar{B}\}\\ &= (1 - \Pr\{B\})\Pr\{A/\bar{B}\}\end{aligned}$$

which yields finally

$$\text{(I)} \qquad \Pr\{A/\bar{B}\} = \frac{\Pr\{A\}}{1 - \Pr\{B\}}.$$

Equation (I) obviously implies the inequality:

$$\text{(II)} \qquad \Pr\{A/\bar{B}\} > \Pr\{A\}.$$

The left-hand side refers to the situation after the refutation of B, the right-hand side to the situation before this refutation. Therefore, we can read (II) as follows: *our confidence in a conjecture can only increase when an incompatible rival conjecture has been exploded.* (Cf. sect. 13.3.)

Yet we can learn more from equation (I). Let us consider $\Pr\{A\}$ as constant and $\Pr\{B\}$ as variable. Let us determine the bounds between which $\Pr\{B\}$ may vary. Of course, $\Pr\{B\}$ can be arbitrarily small. Yet $\Pr\{B\}$ cannot be arbitrarily large; in fact it can never exceed $\Pr\{\bar{A}\}$. If B is correct, $\bar{A}$ is *a fortiori* correct. Since $\Pr\{\bar{A}\}$ is equal to $1 - \Pr\{A\}$

$$0 < \Pr\{B\} \leqq 1 - \Pr\{A\}.$$

We see from equation (I) that

$$\text{as } \Pr\{B\} \text{ increases from } 0 \text{ to } 1 - \Pr\{A\}$$
$$\Pr\{A/\bar{B}\} \text{ increases from } \Pr\{A\} \text{ to } 1.$$

That is, *the more confidence we placed in an incompatible rival of our conjecture, the greater will be the gain of faith in our conjecture when that rival is refuted.*

9. Examining several consequences in succession. We consider now the following important situation: the aim of our research is a certain conjecture A. For the moment, we do not see how we could decide whether A is true or not. Yet we see several consequences $B_1, B_2, B_3, \ldots$ of A:

A implies B_1, A implies B_2, A implies B_3, ... The consequences $B_1, B_2, B_3, \ldots$ are more accessible than A itself and we settle down to examine them one after the other. (This is the typical procedure of the natural sciences: we have no means of examining a general law A in itself, and therefore we examine it by testing several consequences $B_1, B_2, B_3, \ldots$.) We have already examined the consequences B_1, B_2, . . . B_n, and we have succeeded in verifying all of them: $B_1, B_2, \ldots B_n$ turned out to be correct. Now we are testing the next consequence B_{n+1} : how will the outcome affect our confidence in A?

In order to see the answer in the light of the calculus of probability, we start from a general rule of this calculus (see ex. 14.26 (5)):

$$\Pr\{A/H\} \Pr\{B/HA\} = \Pr\{B/H\} \Pr\{A/HB\}.$$

We set $B = B_{n+1}$. Now, since B_{n+1} is a consequence of A,

$$\Pr\{B/HA\} = \Pr\{B_{n+1}/HA\} = 1$$

and so we find that

$$\Pr\{A/H\} = \Pr\{B_{n+1}/H\} \Pr\{A/HB_{n+1}\}.$$

We set $H = B_1 B_2 \ldots B_n$ and obtain the decisive formula:

$$\text{(I)} \qquad \Pr\{A/B_1 \ldots B_n\} = \Pr\{B_{n+1}/B_1 \ldots B_n\} \Pr\{A/B_1 \ldots B_n\, B_{n+1}\}.$$

In order to understand (I) correctly, we have to realize that

$$\Pr\{A/B_1 \ldots B_n\} \text{ and } \Pr\{B_{n+1}/B_1 \ldots B_n\}$$

denote the credibilities of A and B_{n+1}, respectively, *after* $B_1, B_2, \ldots B_n$ have been verified but, of course, *before* B_{n+1} has been verified;

$\Pr\{A/B_1 \ldots B_n\, B_{n+1}\}$ denotes the credibility of A after the verification of its $n + 1$ consequences, $B_1, B_2, \ldots B_n$ and B_{n+1}.

We have to keep these meanings in mind, and then we can read (I) as a precise and pregnant proposition on inductive reasoning.

Let us first focus our attention on $\Pr\{B_{n+1}/B_1 \ldots B_n\}$; the value of this credibility will be in most cases less than 1, and equal to 1 only if the correctness of $B_1, B_2, \ldots B_n$ renders the correctness of B_{n+1} certain, that is, if $B_1, B_2, \ldots$ and B_n jointly imply B_{n+1}. If this is *not* the case, we can derive from (I) the inequality:

$$\text{(II)} \qquad \Pr\{A/B_1 \ldots B_n\} < \Pr\{A/B_1 \ldots B_n\, B_{n+1}\}.$$

That is, *the verification of a new consequence enhances our confidence in the conjecture, unless the new consequence is implied by formerly verified consequences.*

Let us write equation (I) in the form:

$$\text{(III)} \qquad \Pr\{A/B_1 \ldots B_n\, B_{n+1}\} = \frac{\Pr\{A/B_1 \ldots B_n\}}{\Pr\{B_{n+1}/B_1 \ldots B_n\}}.$$

The left-hand side refers to the situation after the confirmation of B_{n+1}; the right-hand side refers to the situation before this confirmation. Let us regard the relation of A to $B_1, B_2, \ldots B_n$ as fixed, but the relation of B_{n+1} to $B_1, B_2, \ldots B_n$ as variable. Then we can read (III) as follows: *the increase in our confidence brought about by the confirmation of a new consequence* (or the weight of the evidence furnished by this confirmation) *varies inversely as the credibility of the new consequence, appraised* (before its confirmation, of course) *in the light of the previously verified consequences.*

We can express this same rule in other words. When we start testing the consequence B_{n+1} of our conjecture A, we face the possibility that B_{n+1} will turn out false in which case A will be exploded. In view of the formerly verified consequences $B_1, B_2, \ldots B_n$ the chance for exploding A by disproving B_{n+1} appears strong when $\Pr\{B_{n+1}/B_1 \ldots B_n\}$ is small. Therefore we can read (III) as follows: *the consequence that, judged in the light of the preceding verifications, stands the better chance of refuting the proposed conjecture, will disclose the stronger inductive evidence if it is confirmed in spite of the forebodings.* Still shorter: "more danger, more honor." If a conjecture escapes the danger of refutation it shall be esteemed in proportion to the risk involved.

From the very beginning of our discussion we have considered the inductive evidence supplied by the successive verification of several consequences of a proposed conjecture. The extreme cases were the most conspicuous. Let us survey them once more (adding just a little color) and let us focus the moment when, having verified the consequences $B_1, B_2, \ldots B_n$ of a conjecture A, we start scrutinizing a new consequence B_{n+1}.

The new consequence under scrutiny, B_{n+1}, may appear "little different" from the formerly verified consequences $B_1, B_2, \ldots B_n$. Such a case is not too exciting. We confidently expect (by analogy, presumably) that B_{n+1} will be verified like the other consequences (that is, $\Pr\{B_{n+1}/B_1 \ldots B_n\}$ is close to 1, its maximum). We scarcely expect that the investigation of B_{n+1} will disclose some very new aspect or that it will upset the conjecture A, but also, when B_{n+1} is finally verified, the gain in evidence for A is not much.

On the other hand, the new consequence under scrutiny, B_{n+1}, may appear as "very different" from the formerly verified consequences $B_1, B_2, \ldots B_n$. Such a case may be exciting. Analogy with $B_1, B_2, \ldots B_n$ gives us little reason to expect that B_{n+1} will be verified ($\Pr\{B_{n+1}/B_1 \ldots B_n\}$ is close to its minimum). We realize that the investigation of B_{n+1} risks upsetting the conjecture A, but it has also a chance to disclose some new aspect, and when B_{n+1} is eventually verified, the gain in evidence for A may be considerable.

The reader should review some of our former examples and discussions. (Cf. sect. 3.1–3.7, ch. VI, sect. 10.1, sect. 12.2, sect. 13.11, and several other passages.) After due comparison, equation (III) of the present section may appear to him as the most concise and precise expression of the principle involved. At any rate, if he can see the bearing of equation (III) on some of our examples, he has taken a good step towards clarifying his ideas about an important subject.

10. On circumstantial evidence. We now consider a situation that we encountered in dealing with reasoning in judicial matters: we are examining a conjecture A. (This conjecture A may be an accusation advanced by the prosecution.) We (the jury) have to find out whether A is true or not. A circumstance B is submitted (by the prosecution) which is so connected with the conjecture A that

B with A is more credible than without A.

In the course of the proceedings this circumstance B is so strongly confirmed that we may regard it as a proven fact. (Perhaps B was not even challenged by the defense.) How does all this affect our belief in A?

Let the calculus of probability answer this question. The essential assumption concerning the connection between A and B is expressed by the inequality

$$\text{(I)} \qquad \Pr\{B/A\} > \Pr\{B/\bar{A}\}.$$

By basic formulas of the theory of probability (cf. ex. 14.26)

$$\Pr\{A\}\Pr\{B/A\} = \Pr\{B\}\Pr\{A/B\},$$

$$\Pr\{B\} = \Pr\{A\}\Pr\{B/A\} + (1 - \Pr\{A\})\Pr\{B/\bar{A}\}.$$

Combining these, we obtain

$$\text{(II)} \qquad \frac{\Pr\{A\}}{\Pr\{A/B\}} = \Pr\{A\} + (1 - \Pr\{A\})\frac{\Pr\{B/\bar{A}\}}{\Pr\{B/A\}}.$$

Using (I), we conclude from (II) that

$$\text{(III)} \qquad \Pr\{A\} < \Pr\{A/B\}.$$

Both sides of this inequality represent the credibility of the conjecture A, the left-hand side before the verification of the circumstance B, the right-hand side after the verification of B. Therefore, inequality (III) expresses a rule: *if a certain circumstance is more credible with a certain conjecture than without it, the proof of that circumstance can only enhance the credibility of that conjecture.* (Cf. sect. 13.13.)

Yet we can learn more from equation (II). Let us regard $\Pr\{A\}$ and $\Pr\{B/A\}$ as constant, but $\Pr\{B/\bar{A}\}$ as variable. Then $\Pr\{A/B\}$ depends on $\Pr\{B/\bar{A}\}$:

as $\Pr\{B/\bar{A}\}$ decreases from $\Pr\{B/A\}$ to 0

$\Pr\{A/B\}$ increases from $\Pr\{A\}$ to 1.

That is, *the less credible a circumstance appears without a certain conjecture, the more will the proof of that circumstance enhance the credibility of that conjecture.* In sect. 13.13, led by the consideration of examples, we came very close to this rule.

Strong judicial evidence results from several proved coincidences all pointing to the same conclusion; see sect. 13.13 (4). If there are several circumstances B_1, B_2, B_3, . . . each of which is more credible with A than without A, and they are successively proved, the evidence for A increases at each step. The amount of additional evidence resulting from a circumstance newly proved depends on various points. A new circumstance that is very different from the circumstances previously examined (a new witness who is visibly independent of the witnesses previously examined) carries particular weight. To express these points, we could develop formulas so related to those introduced in the present section as the formulas developed in sect. 9 are related to those introduced in sect. 6.

EXAMPLES AND COMMENTS ON CHAPTER XV

1. Examine the situation discussed in sect. 13.8 by the Calculus of Probability.

2. Examine the pattern encountered in the solution of ex. 13.8 by the Calculus of Probability. Can you justify it?

3. Reexamine ex. 13.10 by the Calculus of Probability.

4. *Probability and credibility.* The peculiar "non-quantitative" application of the Calculus of Probability, discussed in sect. 5–10, was designed to elucidate certain patterns of plausible reasoning. These patterns were mainly suggested by heuristic reasoning about mathematical conjectures. Can we apply the Calculus of Probability to examples of other kind in the same manner?

(1) Let A_n denote the conjecture that the fair die that I am about to roll will show n spots ($n = 1, 2, \ldots 6$). These conjectures $A_1, A_2, \ldots A_6$ are of a kind that we did not particularly wish to examine in sect. 5–10. Yet let us try to treat them in the same manner: we consider their credibilities $\Pr\{A_1\}, \Pr\{A_2\}, \ldots \Pr\{A_6\}$ and apply the Calculus of Probability to them. As $A_1, A_2, \ldots A_6$ are mutually exclusive and exhaust all the possibilities,

$$\Pr\{A_1\} + \Pr\{A_2\} + \ldots + \Pr\{A_6\} = 1.$$

Since the die with which we are concerned is fair, its faces are interchangeable, none of the six faces is preferable to any other, and so we are compelled to assume that

$$\Pr\{A_1\} = \Pr\{A_2\} = \ldots = \Pr\{A_6\}.$$

Combining this with the foregoing equation, we obtain that

$$\Pr\{A_1\} = \Pr\{A_2\} = \ldots = \Pr\{A_6\} = 1/6,$$

and so we are led to attribute definite numerical values to the credibilities considered. Mr. Anybody (our friend from sect. 5) would be driven to the same conclusion, I think.

The credibility of the conjecture A_1 turned out to have the same numerical value as the probability of the event that a fair die shows one spot. Yet this is not surprising at all: we admitted the same rules and assumed the same interchangeability (or symmetry) in computing credibilities and probabilities. (The reader should not forget, of course, that credibility and probability are quite differently defined.)

(2) The foregoing argument applies to many other cases.

By three straight lines passing through its center, the surface of a circle is divided into six equal sectors. A raindrop is about to fall on the circular surface. Let us call A_1 the assertion that the raindrop will fall on the first sector, A_2 the assertion that it will fall on the second sector, and so on. This situation is essentially the same as that discussed under (1) and the result is, of course, also the same: each of the six assertions $A_1, A_2, \ldots A_6$ has the same credibility 1/6.

The generalization is obvious: we need not stick to the number 6 or to raindrops. We can divide the circle into n sectors and consider some other kind of random event. We can also pass from the area of the circle to its periphery and so we arrive at the following situation: a point will be chosen on the periphery of a circle by some random agency that has no preference whatsoever for any particular points of this periphery. Somebody conjectures that the point will lie on a certain arc. The credibility of this conjecture is the ratio of the length of the arc to the length of the whole periphery.

From the circle we can pass to the sphere. Let us assume, for the sake of simplicity, that the Earth is a perfect sphere and that the meteorites striking the Earth have no preference for any particular direction. Somebody conjectures that the next meteorite falling on the Earth will strike a certain region. What is the credibility of this conjecture? The mathematical treatment may be more complicated if we carry it through in minute detail, but the result is just as intuitive as for the circle: the desired credibility is the ratio of the area of that region to the area of the whole spherical surface.

We need not enter upon further analogous, or more general, situations. Let us look, however, at two particular cases.

The credibility of the conjecture that a point chosen at random on the periphery of a circle will lie within a distance of one degree from a given point of this periphery is 1/180.

The credibility of the conjecture that the next meteorite falling on the Earth will strike within a distance of one degree from the center of New York City is 0.00007615. (This is the numerical ratio of a small spherical cap to the whole spherical surface.)

The circle and the sphere possess high symmetry: a suitable rotation that does not change the position of the figure as a whole can transport a point of the figure from any given position on the figure to any other given position. This symmetry is not enough in itself to justify the preceding results. In deriving them we have to *assume* "physical" symmetry, the interchangeability of any two points of the geometric figure in relation to the physical agency under consideration.

The computed credibilities cannot claim much novelty: they coincide with corresponding probabilities, known for a long time, and successfully applied to a variety of mass phenomena.

The computed credibilities may appear as incapable of interfering with the credibilities with which we are primarily concerned. Of this, however, we should not feel too sure.

(3) Kepler knew only six planets revolving around the sun and even devised a geometric argument why there *should* be exactly six planets; see sect. 11.5. Yet the telescopes did not listen to his argument. In 1781, about 150 years after Kepler's death, the astronomer Herschel observed a slowly moving star and supposed it to be a comet; but it turned out to be a seventh planet (Uranus) revolving beyond the orbit of Saturn. In the years 1801–1806 four small planets (Ceres, Pallas, Juno, and Vesta) revolving between the orbits of Mars and Jupiter were similarly discovered. (Hundreds of such minor planets have been discovered later.)

On the basis of Newton's theory, the astronomers tried to compute the motions of these planets. They did not succeed very well with the planet Uranus; the differences between theory and observation seemed to exceed the admissible limits of error. Some astronomers suspected that these deviations may be due to the attraction of a planet revolving beyond Uranus' orbit, and the French astronomer Leverrier investigated this conjecture more thoroughly than his colleagues. Examining the various explanations proposed, he found that there is just one that could account for the observed irregularities in Uranus' motion: the existence of an ultra-Uranian planet. He tried to compute the orbit of such a hypothetical planet from the irregularities of Uranus. Finally Leverrier succeeded in assigning a definite position in the sky to the hypothetical planet. He wrote about it to another astronomer whose observatory was the best equipped to examine that portion of the sky. The letter arrived on the 23rd of September, 1846 and in the evening of the same day a new planet was found within one degree of the spot indicated by Leverrier. It was a large ultra-Uranian planet that had approximately the mass and orbit predicted by Leverrier.

(4) The theory that renders possible such an extraordinary prediction must be a wonderful theory. This may be our first impression. Let us try to clarify this impression by the Calculus of Probability.

T stands for the theory that underlies the astronomical computations: it is Newton's theory, consisting of his laws of mechanics and his law of gravitational attraction.

N stands for Leverrier's assertion: at the date given, there will be a new planet, having such and such a mass and such and such an orbit (approximately), in the neighborhood of such and such a point of the sky. More precisely, let N stand for that part of Leverrier's assertion that has been *verified* by subsequent observations.

$\Pr\{T\}$ denotes the degree of confidence placed in the theory T on the basis of all facts known by the astronomers before the date of the discovery of the new planet.

$\Pr\{T/N\}$ denotes the degree of confidence due to the same theory T when the verification of Leverrier's prediction N has been added to the facts known before.

We think that the verification of Leverrier's prediction adds to the credit of the theory T, and so we suspect that $\Pr\{T/N\}$ is greater than $\Pr\{T\}$. In fact, we find (by ex. 14.26 (2)) that

$$\frac{\Pr\{T/N\}}{\Pr\{T\}} = \frac{\Pr\{N/T\}}{\Pr\{N\}}.$$

$\Pr\{N\}$ is the credibility of Leverrier's assertion N "in itself," that is, without reference to the truth or falsity of the theory T; the underlying state of knowledge is the same as assumed in the evaluation of $\Pr\{T\}$. This $\Pr\{N\}$ must be extremely small: if we ignore Newton's theory T what ground could we have to suspect that there is an ultra-Uranian planet of such and such precise properties so near to a given point of the sky?

$\Pr\{N/T\}$ is the credibility of Leverrier's assertion N in the light of Leverrier's computation based on the theory T, and so it is a concept very different from $\Pr\{N\}$. Perhaps Leverrier did not show quite conclusively that the existence of an ultra-Uranian planet with such and such properties is the only explanation compatible with the theory T. Yet he came pretty near to showing just this, and so $\Pr\{N/T\}$ cannot be too far from certainty.

To sum up, we may regard $\Pr\{N/T\}$ as a sizable fraction, not very far from 1, and $\Pr\{N\}$ as a very small fraction, approaching 0. Yet if we think so, the ratio $\Pr\{N/T\}/\Pr\{N\}$ appears as very large, and so must appear the ratio $\Pr\{T/N\}/\Pr\{T\}$ which has the same value as the foregoing, by the above equation. Therefore, $\Pr\{T/N\}$, our confidence in the theory T after the verification of Leverrier's prediction appears as very much greater than $\Pr\{T\}$, our confidence in the same theory before this verification.

(5) The foregoing consideration remains within the limits to which we cautiously restricted ourselves in sect. 5. Yet let us discard caution for a moment and indulge in some adventurous rough estimates.

Leverrier's prediction N covers many details of which we pick out just one: the new planet will be near such and such a spot of the sky. It was found, in fact, within one degree from the indicated point (at a distance 52′). Yet the probability that a point chosen at random on the sphere should be within one degree from a preassigned spot can be computed under simple assumptions, as we stated above, under (2). We find that $\Pr\{N\}$ is much smaller than 0.00007615. We can scarcely regard $\Pr\{N/T\}$ as exactly equal to 1, yet we may surmise that, in setting

$$\Pr\{N\} = 0.00007615, \qquad \Pr\{N/T\} = 1,$$

we overestimate much more the first, than the second, credibility. And so we arrive at the inequality

$$\frac{\Pr\{T/N\}}{\Pr\{T\}} > \frac{1}{0.00007615} = 13131.$$

Of course, such an estimate is questionable. There may be reasons of analogy, quite independent of Newton's theory T, suggesting that a new planet has more chance to be near the plane of the Earth's orbit than far from it. If we think so, we should substitute for 0.00007615 a greater fraction but one that is still less than

$$1/180 = 0.005556;$$

cf. under (2).

Such estimates could be argued endlessly. For example, since $\Pr\{T/N\}$ is certainly less than 1, the inequality proposed implies that

$$\Pr\{T\} < 0.00007615.$$

We may be tempted to regard this as a refutation of the proposed inequality, as a "reductio ad absurdum." In fact, Newton's theory T could be regarded as firmly established in 1846, even before the discovery of Neptune, and so it could appear as absurd to attribute to T such a low credibility. I do not think, however, that we are obliged to regard a credibility 10^{-5} as low in such a case: we could think that logical certainty, to which we ascribe the credibility 1, is incomparably more than the reliance that we can place in the best established inductive generalization, the credibility of which we could even consider as infinitesimal; cf. ex. 8.

After all this, we may find it safer to return to the standpoint of sect. 5–10, to which we essentially adhered under (4). Represent to yourself qualitatively how a change in this or that component of the situation would influence your confidence, but do not commit yourself to any quantitative estimate.

5. *Likelihood and credibility.* We may have a conjecture about probabilities. For example, you may conjecture that the die that you hold in your hands is ideally fair, that is, each face has the same probability 1/6. Of course, this conjecture is hard to believe. Or you may conjecture that each face of that die has a probability between 0.16 and 0.17, which would be more credible. A conjecture about probabilities is a statistical hypothesis. It often happens that we have only two obvious rival conjectures: a "physical" conjecture P and a statistical hypothesis H; cf. sect. 14.9 (7) and ex. 14.33. In such a case we may be seriously concerned with $\Pr\{H\}$, the credibility of the statistical hypothesis H.

A statistical hypothesis H is appropriately tested by statistical observation. Let E denote the prediction that the statistical observation will yield such and such a result. Let us consider the credibility $\Pr\{E/H\}$ and let us assume that this *credibility has a numerical value, equal to the probability* that an event of the kind predicted by E will happen, computed on the basis of the statistical hypothesis H. As we have seen in ex. 4, certain (quite natural) assumptions of interchangeability, or symmetry (included in the statistical hypothesis H) may even oblige us to equate credibility with probability.

This credibility, or probability, $\Pr\{E/H\}$ can be viewed from two different standpoints as we have discussed in sect. 14.7 (5); cf. also sect. 14.8 (5). On the one hand $\Pr\{E/H\}$ is the probability of an event of the kind predicted by

E, computed on the basis of the statistical hypothesis H. On the other hand, if such an event actually happens and is observed, we are inclined to think that H is the less likely the less the numerical value of $\Pr\{E/H\}$ is, and for this reason we call $\Pr\{E/H\}$ the *likelihood* of the statistical hypothesis H, judged in view of the fact that the event predicted by E has actually happened. Cf. sect. 14.7 (5), also sect. 14.8 (5).

Now it follows from ex. 14.26 (2) that

$$\Pr\{H/E\} = \frac{\Pr\{E/H\}\Pr\{H\}}{\Pr\{E\}}.$$

In this equation, $\Pr\{E/H\}$ is not only a credibility, but also a probability, and has a definite numerical value. Yet $\Pr\{H/E\}$, $\Pr\{H\}$, $\Pr\{E\}$ are only credibilities and are not supposed to possess definite numerical values.[4] Especially both $\Pr\{H\}$ and $\Pr\{H/E\}$ denote the credibility of the same statistical hypothesis H, but the first before, and the second after, the observation of the event predicted by E. Let us restate the above equation in a less conventional form, emphasizing one of the aspects of $\Pr\{E/H\}$:

$$\text{Credibility after event} = \frac{\text{Likelihood} \times \text{Credibility before event}}{\text{Credibility of event}}.$$

Having observed the event predicted by E, we face a decision: should we reject the statistical hypothesis H, and accept the rival physical conjecture P, or what should we do? Our decision should be based on the latest information, and so on $\Pr\{H/E\}$, the credibility of the statistical hypothesis after the observation of the event. Of this credibility $\Pr\{H/E\}$ the likelihood $\Pr\{E/H\}$ is a factor: it is the most important factor, perhaps, because it possesses a numerical value, computable by a clear and familiar procedure, but it is still just a factor, not the full expression of the credibility.

The likelihood is an important indication, but not everything. The statistician may wisely restrict himself to the computation of the likelihood, but the statistician's customer may act unwisely if he neglects the other factors. He should carefully weigh $\Pr\{H\}$, the credibility of the statistical hypothesis H before the event: we mean, in fact, this $\Pr\{H\}$ when we talk about the "appropriateness," or "realism," of H. Cf. ex. 14.33.

6. *Laplace's attempt to link induction with probability.* A bag contains black and white balls in unknown proportions; m white and n black balls have been successively drawn and replaced. What is the probability that $m' + n'$ subsequent drawings will yield m' white and n' black balls?

A particular case of this apparently harmless problem about the bag and the balls can be interpreted as the fundamental problem about inductive inference reduced to its simplest expression. In fact, let us consider the case

[4] This is the usual situation. Only exceptionally is the statistician in a position to ascribe a numerical value to $\Pr\{H\}$.

where $n = n' = 0$. We drew m balls from the bag of unknown composition, and all balls drawn turned out to be white. We may assimilate this situation to the naturalist's situation who tested m consequences of a conjecture and found all m observations concordant with the conjecture. The naturalist plans more observations. What is the probability that his next m' observations will also turn out to be concordant with the conjecture? This question may be construed as the particular case $n = n' = 0$ of the proposed problem about the bag and the balls.

In this problem there is an obscure and puzzling point: the proportion of the black balls to the white balls in the bag is unknown. Yet in view of the intended interpretation, this point appears essential: the naturalist cannot know the "inner workings" of nature. He knows only what he has observed, and we know only that the bag yielded, up to this moment, so and so many white balls in so and so many drawings.

Common sense would suggest that we cannot compute the required probability if nothing is known about the composition of the bag; a problem without sufficient data is insoluble. Yet Laplace gave a solution—a controversial solution. How did he manage to arrive at a solution at all?

Laplace introduces a new principle to compensate for the lack of data; this principle is controversial. "When the probability of a simple event is unknown, we may suppose all possible values of this probability between 0 and 1 as equally likely," says Laplace.[5] "This is the equal distribution of ignorance," sneer his opponents.

Once Laplace's controversial principle is admitted, the derivation of the solution is straightforward; we need not consider it here. Its result is: if m drawings yielded only white balls, the probability that m' subsequent drawings should also yield only white balls is

$$\frac{m+1}{m+m'+1}.$$

Let us call this statement the "General Rule of Succession." The best known particular case is concerned with $m' = 1$: if m drawings yielded only white balls, the probability that also the next drawing should yield a white ball is

$$\frac{m+1}{m+2}.$$

Let us call this the "Special Rule of Succession."[6]

Are these rules acceptable if we interpret "white balls" as "concordant observations of the same nature" and "probability" as "degree of reasonable confidence"? This question is to the point and we shall discuss it.

---

[5] *Oeuvres complètes*, vol. 7, p. XVII.

[6] This differs somewhat from the usual terminology. Cf. J. M. Keynes, *A treatise on probability*, p. 372–383.

(1) Let us reconsider our first example of inductive reasoning. Goldbach's conjecture asserts that, from $6 = 3 + 3$ onward, any even integer is the sum of two odd primes. The table in sect. 1.3 verifies this conjecture up to 30. Having verified it up to 30, we expect with more or less confidence that it will also be verified in the next case, for 32. The Special Rule of Succession may be interpreted to mean that, having verified Goldbach's conjecture in the first m cases, we are entitled to expect its verification in the next case with the probability

$$\frac{m+1}{m+2} = 1 - \frac{1}{m+2}.$$

Let us realize what this means. As m increases the probability also increases: in fact, the more cases have been verified in the past, the more confidently we expect the conjecture to be verified in the next case. If m tends to ∞, the probability tends to 1: we could hope to approach certainty closer and closer by collecting more and more verifications. Let us now consider the difference of two probabilities, one corresponding to $m + 1$, the other to m, previous verifications:

$$\frac{m+2}{m+3} - \frac{m+1}{m+2} = \frac{1}{(m+2)(m+3)}.$$

This difference decreases as m increases: it is true that each new verification adds to our confidence, but it adds less and less as it comes after more and more previous similar verifications. (The similarity of the verifications is essential at this point; cf. sect. 12.2.)

Let us take up now the General Rule of Succession. It could be interpreted to mean this: having verified Goldbach's conjecture in the first m cases, we are entitled to expect its verification in the next m' cases with the probability

$$\frac{m+1}{m+m'+1}.$$

If we keep m fixed, but let m' increase, this probability decreases: in fact, the further we try to predict the future on the basis of past observation, the less confidently we can predict it. If m' increases indefinitely, the probability tends to 0. In fact, the verification for all values of m' would mean that Goldbach's conjecture is true. Obviously, on the basis of a given number m of observations we cannot assert that the conjecture is true. The rule seems to imply a stronger statement: on the basis of m observations we cannot even attribute a probability different from 0 to Goldbach's conjecture, and such a strong statement may point in the right direction.

(2) Up to this point the Rule of Succession looks rather respectable. Yet let us look at it more concretely. Let us attribute numerical values to m and let us not neglect everyday situations. It will be enough to consider the Special Rule of Succession.

I tested the even numbers 6, 8, 10, . . . 24, and found that each of them is a sum of two odd primes. The Rule says that I should expect with the probability 11/12 that also 26 is a sum of two odd primes.

In a foreign city where I scarcely understood the language, I ate in a restaurant with strong misgivings. Yet after 10 meals taken there I felt no ill effects and so I went quite confidently to the restaurant the eleventh time. The Rule said that the probability that I would not be poisoned by my next meal was 11/12.

A boy is 10 years old today. The Rule says that, having lived 10 years, he has the probability 11/12 to live one more year. The boy's grandfather attained 70. The Rule says that he has the probability 71/72 to live one more year.

These applications look silly, but none is sillier than the following due to Laplace himself. "Assume," he says, "that history goes back 5,000 years, that is, 1,826,213 days. The sun rose each day, and so you can bet 1,826,214 against 1 that the sun will rise again tomorrow."[7] I would certainly be careful not to offer such a bet to a Norwegian colleague who could arrange air transportation for both of us to some place within the Arctic circle.

Yet the rule can beat even this absurdity. Let us apply it to the case $m = 0$: the Rule's derivation is as valid for this case as for any other case. Yet for $m = 0$ the Rule asserts that any conjecture without any verification has the probability 1/2. Anybody can invent examples to show that such an assertion is monstrous. (By the way, it is also self-contradictory.)

(3) Our discussion was long. In more guarded language it would have been still longer, and it could be prolonged, but here is the long and the short of it: the Rule of Succession may look wise if we avoid numerical values, but it certainly looks foolish if we get down to numerical values. Perhaps, this points to a moral: in applying the Calculus of Probability to plausible reasoning, avoid numerical values on principle. At any rate, this is the standpoint advocated in this chapter.

7. *Why not quantitative?* This chapter advances a thesis: The Calculus of Probability should be applied to plausible reasoning, but only qualitatively. But there is a strong temptation to apply it quantitatively, and so we have to examine a few more relevant points.

(1) *Non-comparable.* There is some evidence that Goldbach's conjecture concerning the sum of two odd primes is correct; see sect. 1.2–1.3. There is some evidence that Norsemen landed on the American mainland a few hundred years before Columbus. Which evidence is stronger?

This seems to be a very silly question indeed. What could be the purpose of comparing two such disparate cases? And who should compare them? To judge the evidence competently you should be an expert. In one case the evidence should be judged by a mathematician expert in the Theory of

[7] *Oeuvres complètes,* vol. 7, p. XVII.

Numbers. In the other case the evidence should be judged by a historian expert in Old Icelandic. There is scarcely a person who would be expert in both.

However this may be, our apparently silly question points out a possibility. It could be that there is no reasonable decision, no reasonable way to say which evidence is stronger than the other. This possibility is so important that it deserves a name. If there is no reasonable way to decide which evidence is stronger, E_1 or E_2, let us call E_1 *non-comparable* with E_2. We could find in the foregoing chapters several examples that suggest more clearly and more convincingly than the example from which we started here that one evidence may not be comparable with another. See sect. 4.8.

(2) *Comparable.* After having pointed out the possibility that evidences may be non-comparable, let us survey now a few cases in which evidences are obviously comparable.

Let E_1 denote the evidence for Goldbach's conjecture (sect. 1.2) resulting from its verification for all even numbers up to 1,000. Let E_2 denote the evidence for the same conjecture resulting from its verification up to 2,000. Obviously, the evidence E_2 is stronger than the evidence E_1.

Now, let us change the notation, and let E_1 denote the present evidence for the landing of Norsemen on the American continent before Columbus. Let E_2 denote what this evidence would grow into, if somebody discovered, say, a burial place somewhere on the coast of Labrador containing shields and swords similar to those preserved elsewhere from the Viking age. Obviously, the evidence E_2 would be stronger than the evidence E_1.

Let us consider one more, somewhat subtler, case. Let now E_1 denote the evidence for a conjecture A resulting from the verification of one of its consequences B. After the consequence B has been verified, somebody remarks that B is very improbable in itself. (This remark could be quite precise; the probability of B, computed on the basis of a simple and apparently appropriate statistical hypothesis, could be very low.) This remark changes the evidence for the conjecture A into E_2. The evidence E_2 is stronger than the evidence E_1. (We have said this already, perhaps a trifle less sharply, in sect. 12.3.)

In all three examples, we obtain the evidence E_2 from the evidence E_1 by adding some relevant observation. Yet, if there is no such simple relation between E_1 and E_2, how could we decide which one is stronger? This question brings even closer the possibility that evidences may be non-comparable.

(3) *Comparable, but still not quantitative.* In the foregoing, subsection (2), we have seen cases in which an evidence E_2 may be reasonably held stronger than an evidence E_1. Yet *how much stronger?* It seems to me that there is no reasonable answer to this question in any of the foregoing cases. And so we still remain on a qualitative level.

(4) *How would it look?* In sect. 4 we started taking the symbol $\Pr\{A\}$ to denote the credibility of a conjecture A. In the subsequent sections of this chapter we tried to get along without giving any determinate numerical value to $\Pr\{A\}$: herein consists the "qualitative" standpoint that this book advocates. The "quantitative" standpoint would consist in giving to $\Pr\{A\}$ a definite numerical value whenever the Calculus of Probability is applied to plausible reasoning about the conjecture A.

The burden of proof falls squarely on those who champion a quantitative application of the Calculus of Probability to plausible reasoning. All that they have to do is to produce a class of non-trivial conjectures A for which the credibility $\Pr\{A\}$ can be computed by a clear method that leads to acceptable results at least in some cases.[8]

Nobody has yet proposed a clear and convincing method for computing credibilities in non-trivial cases, and if we visualize concrete situations in which a sound estimate of credibilities is important (as we have done), we can easily perceive that any attribution of definite numerical values to credibilities is in great danger of looking foolish.

Credibilities that have definite numerical values are comparable: two numbers are either equal, or one of them is greater than the other. Yet, after the discussion under (1) and (2), we may find it hard to accept that any two conjectures should be comparable in credibility. Take the two conjectures with which we began our discussion: Goldbach's conjecture and that historical conjecture about the discovery of America. If we attributed numerical values to their credibilities, the strength of the evidence speaking for one could be compared with that for the other: yet such a comparison looks futile and foolish.

(5) *Would it be worthwhile?* There is another point to consider which is independent of the foregoing discussion. The weight of a plausible argument may be extremely important, but such importance is provisional, ephemeral, transient: would it be worthwhile to fasten a numerical value on something so transitory?

What is the credibility of Newton's law of gravitation, judged in the light of the facts collected in the first edition of the Principia? Let us imagine for a moment that there is a method to evaluate numerically such a credibility. Yet we should not imagine for a moment that the evaluation could be easy: in view of the complexity of the facts and their interrelations, the evaluation must be delicate and the numerical computation long. Would it be worthwhile to undertake it? Perhaps, for us, in view of the historical and philosophical importance of Newton's discovery. Yet scarcely for Newton and his contemporaries: instead of computing the credibility of the theory, they could have, with the same effort, changed the credibility by developing

[8] Attributing a numerical value to a credibility on the basis of some assumption of interchangeability or symmetry (as we did in ex. 4) should be regarded here as trivial: something more, and more novel, is required to justify quantitative credibilities.

the theory, and multiplying the observations. It looks preposterous to devote ten years to the computation of a degree of credibility that is valid only a second.

8. *Infinitesimal credibilities?* A new particular case of some number-theoretic conjecture (such as Goldbach's conjecture; cf. sect. 1.2–1.3) has been verified. Such verification must be considered as increasing the weight of the evidence, or the credibility of the conjecture. Yet no amount of such verifications can prove the conjecture. We may even feel that no amount of such verifications can bring the conjecture measurably nearer to a proof. (Cf. ex. 4 (5), ex. 6 (1).) Such feelings may suggest the introduction of infinitesimals into the Calculus of Probability.

Infinitesimals can be quite clearly handled by modern mathematics. We consider "quantities" $a, b, \ldots$ represented by "formal power series":

$$a = a_0 + a_1\varepsilon + a_2\varepsilon^2 + a_3\varepsilon^3 + \ldots ;$$

$a_0, a_1, a_2, \ldots$ are real numbers, ε is an indeterminate, no attention is paid to convergence. There is an algebra of such quantities; there are rules (familiar from the theory of convergent power series) according to which such formal power series can be added, subtracted, and multiplied; even division by a can be performed if $a_0 \neq 0$. We call a the "quantity zero" or 0 when $a_0, a_1, a_2, \ldots$ all vanish. Two quantities are equal if their difference is 0. The quantity a reduces to the number a_0 when $a_1, a_2, a_3, \ldots$ all vanish.

We say that a is *positive* if the *first* non-vanishing number in the sequence $a_0, a_1, a_2, \ldots$ is positive. From this definition, we easily derive two basic rules:

Either a is 0 or a is positive or $-a$ is positive, and these three possibilities are mutually exclusive.

The sum and product of two positive quantities are positive.

We say that $a > b$ if $a - b$ is positive. From these definitions it follows that any positive *number* is greater than ε, (that is, the formal power series $0 + \varepsilon + 0 \cdot \varepsilon^2 + 0 \cdot \varepsilon^3 + \ldots$) although ε is positive. And so ε is an "actual" infinitesimal.

We may contemplate now some form of the Calculus of Probability in which the probabilities (or credibilities) are not necessarily numbers, but quantities a of the kind just introduced, subject to the condition $0 \leqq a \leqq 1$. In such a calculus the credibility of Goldbach's conjecture after 1,000,000, or any other number, of verifications could still be infinitesimal, that is, a quantity a with $a_0 = 0$.

I refrain from comments on the prospects of such a plan. At any rate, it discloses a possibility of viewing non-numerical credibilities from another angle.

9. *Rules of admissibility.* I wish the reader to draw his own conclusions and to form his own opinion. Therefore I postponed the expression of my opinion on a crucial question to this last comment of the chapter. I mean the question suggested by the first section of the chapter: Has plausible reasoning rules of any kind?

It is rather obvious that plausible reasoning has no rules of the same kind as demonstrative reasoning. A demonstration has been proposed. If it is presented in sufficiently small steps, the validity of each step can be tested by a rule of formal logic. If all steps conform to the rules, the demonstration is valid, but it is invalid if there is a step violating the rules. Thus, the rules of demonstrative logic are decisive: they can decide whether a proposed demonstrative argument is binding or not. The patterns of plausible reasoning that we have collected cannot achieve such a thing. A plausible argument has been proposed. Each step of it intends to render a certain conjecture more credible and does so following some accepted pattern. Having followed each step of the whole argument, you are not bound to trust the conjecture to any definite degree.

Yet there are rules of different kinds. Logical rules are very different from legal rules. A law court should listen to all parties concerned, but it should not listen to irrelevancies. Therefore a law court should have powers to exclude irrelevant matters from its proceedings, and such powers are regulated by *rules of admissibility*. Without some rules of this kind, there could be no orderly administration of justice: the court could not restrict an unscrupulous counsel who, by unfair or irrelevant questions, could wear down adverse witnesses, the opposition, the jury, and the judge, or drag out the proceedings indefinitely.

The patterns of plausible reasoning collected in the foregoing can be regarded as *rules of admissibility in scientific discussion*. You are by no means obliged to give a definite degree of credence to a conjecture if some of its consequences have been verified. Yet if the conjecture is discussed, it is certainly admissible to mention such verifications and it is fair and reasonable to listen to them. Our patterns register various points concerning such verifications that could reasonably influence the weight of the evidence (as analogy, or lack of analogy, with former verifications, etc.). It is fair and reasonable to admit the discussion of such points too. In collecting these patterns the author's intention was to list those general points that, *according to the usage of good scientists, are admissible in a scientific discussion*, with a view to reasonably influencing the credibility of the conjecture discussed.

In a trial by jury, the powers of the court are divided between the jury and the presiding judge. This division of powers (as conceived by certain legal authorities and accepted to some extent by the judicial practice of certain states and countries) is of great interest for us. The jury and the judge have different functions, they decide different questions. Questions concerning the admissibility of evidence are answered by the judge, questions concerning the credibility of the evidence admitted are answered by the jury. It is for the judge to decide which evidence deserves consideration by the jury.

It is for the jury to determine whether the submitted evidence is of sufficient weight. In deciding which matters are, and which are not, worthy of consideration by the jury, the judge has to know and to respect precedent, the court's usage, and the formulated rules of admissibility. In weighing the evidence submitted, the juror, who had possibly no legal training at all, has to rely on his own natural lights.

In short, the powers of the court are so divided between the judge and the jury, as the power of judging a proposed conjecture is divided in each of us between impersonal rules and personal good sense. The judge plays the role of the rules, the jury that of your personal discernment. It is for the impersonal rules of plausible reasoning to decide which kind of evidence deserves consideration. Yet it is for your personal good sense to decide whether the particular piece of evidence just submitted has sufficient weight or not.

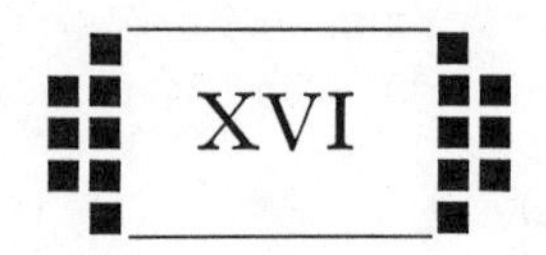

XVI

PLAUSIBLE REASONING IN INVENTION AND INSTRUCTION

Words consist of letters of the alphabet, sentences of words which can be found in the dictionary, and books of sentences which may be found also in other authors. Yet if the things I say are consistent and so connected that they follow from each other, you can as well blame me for having borrowed my sentences from others as for having borrowed my words from the dictionary.
—DESCARTES[1]

1. Object of the present chapter. The examples in the first part of this work and the discussions in the foregoing chapters of the second part elucidated somewhat, I hope, the role of plausible reasoning in the discovery of mathematical facts. Yet the mathematician does not only guess; he also has problems to solve, and he has to prove the facts that he guessed. What is the role of plausible reasoning in the discovery of the solution or in the invention of the proof? This is the question to be discussed in the present chapter. And, by the way, this is the question that attracted the author who, primarily concerned with the methods of problem-solving, was eventually led to the subject of the present book.

The subject of plausible reasoning is subtle and elusive and so is the subject of methods of solution. It was perhaps appropriate to defer the question that combines two such delicate subjects till the last chapter. The following treatment will be brief; the main purpose is to point out the connection with the matters previously discussed. A more ample treatment would fit into another book on methods of problem-solving.

2. The story of a little discovery. The solution of any simple but not merely routine mathematical problem may bring you some of the tension and the triumph of a discovery. Let us look at the following example: *Construct a quadrilateral, being given a, b, c, and d, its four sides, and ε, the angle included by the opposite sides a and c.*

[1] *Oeuvres*, edited by Adam and Tannery, vol. 10, 1908, p. 204.

The data of the problem are exhibited in fig. 16.1: four lines and one angle, fragments of a figure torn to pieces that we should reassemble to satisfy all the requirements laid down in the problem.

It is understood that the sides a, b, c, and d follow each other in this order around the desired quadrilateral so that a is opposite to c, and b to d. The angle ε, included by the opposite sides a and c is *not* one of the four angles of the quadrilateral.

Let us ask a few of the usual questions that may bring the problem closer to us.

Are the data sufficient to determine the unknown? The four sides alone would be obviously insufficient to determine the quadrilateral: four sticks attached by flexible joints at their respective ends form an articulated quadrilateral which is movable, deformable, not rigid, has no determinate shape. Yet if one of its four angles is fixed, the articulated quadrilateral cannot move any more: the quadrilateral is determined by its four sides and one of its angles. We may guess that it is also determined by four sides and some other angle, and so the data of our problem appear sufficient.

Draw a figure. We draw fig. 16.2 which displays all five data so assembled as they should be assembled according to the conditions of the proposed problem. Presumably, we have to *use all the data.*

It can happen that we get stuck at this point and no useful idea occurs to us for a while. In fact, fig. 16.2 looks clumsy. The sides a, b, c, d are at their proper places, of course, but the situation of the angle ε appears unfortunate. This angle is one of our data, we have to use it. Yet how can we use it if it is located so far away, at such an unusual place?

An experienced problem-solver would try to *redraw* the figure: he would try to place that angle ε somewhere else. He may hit so upon fig. 16.3 where the angle ε is between the side a and a parallel to the side c drawn through an endpoint of a. Fig. 16.3 looks more promising than the obvious fig. 16.2.

Why does fig. 16.3 look promising? Even good students who feel pretty sure that fig. 16.3 is promising may be unable to answer this question clearly.

"It looks good to me."

"The data are more tightly fitted together."

Only a student who is exceptionally talented or experienced will be able to give a full explanation: "In fig. 16.2 the angle is located in a triangle. Yet this triangle is not suitable for construction: only two data are known, ε and b. As located in fig. 16.3, the angle ε has more chance to be fitted into a suitable triangle. This is desirable, since, usually, this kind of construction is reduced to the construction of a triangle from suitable data."

The general idea behind the last answer seems to be: any feature in which the present situation recognizably agrees with successful past situations seems promising.

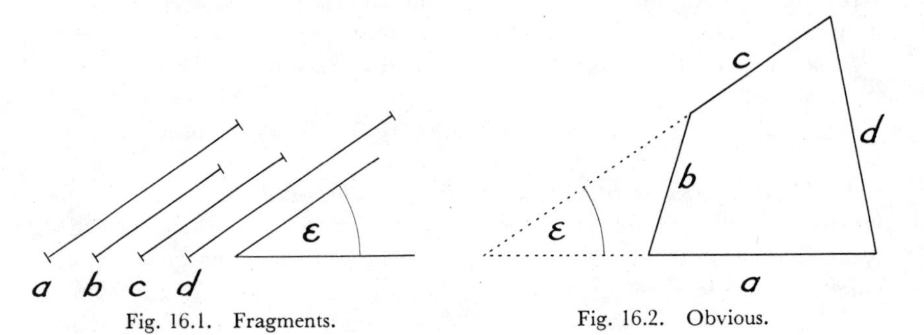

Fig. 16.1. Fragments.

Fig. 16.2. Obvious.

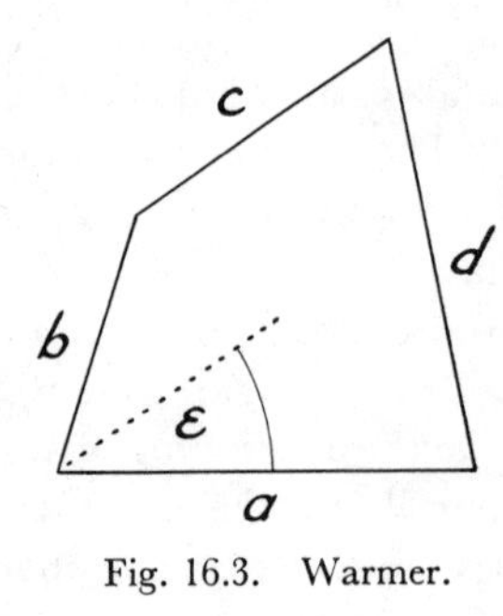

Fig. 16.3. Warmer.

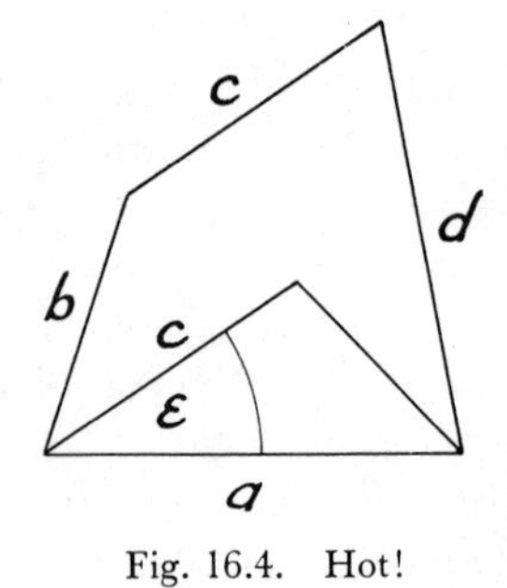

Fig. 16.4. Hot!

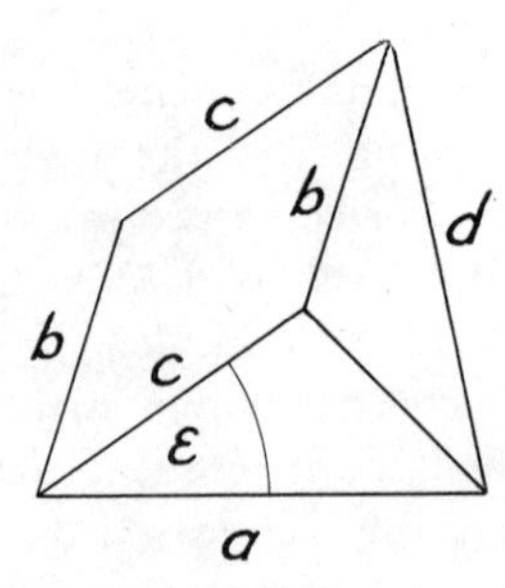

Fig. 16.5. Done!

However this may be, fig. 16.3 lives up to our expectations. In fact, the angle ε does fit into a triangle that we can easily construct (from the sides a and c and the included angle ε, see fig. 16.4). When this triangle has been completed, the solution is quite close. In fact, to the triangle just completed another triangle is attached that we can construct (from sides b, d, and a previously constructed side, see fig. 16.5). Having drawn both triangles, we complete the required construction by drawing the parallelogram with sides b and c.

For most problem-solvers the decisive step in the foregoing solution, represented by a succession of five figures, is the step from fig. 16.2 to fig. 16.3; a longer period of hesitation may precede this step. Yet, once the angle ε is favorably located, the progress of the solution from fig. 16.3 through fig. 16.4 to the final fig. 16.5 may be quite rapid.

The foregoing solution may reveal one or two points which play a rôle also in more momentous discoveries.

3. The process of solution. Solving a problem is an extremely complex process. No description or theory of this process can exhaust its manifold aspects, any description or theory of it is bound to be incomplete, schematic, highly simplified. I wish to point out the place of plausible reasoning in this complex process, and I shall choose the simplest description I am able to find in which this place can be recognizably located. And even the beginning of such a simple description will suffice here.

(1) *Setting a problem to yourself.* A problem becomes a problem for you when you propose it to yourself. A problem is not yet your problem just because you are supposed to solve it in an examination. If you wish that somebody would come and tell you the answer, I suspect that you did not yet set that problem to yourself in earnest. But if you are anxious to find the answer yourself, by your own means, then you have made the problem really yours, you are serious about it.

Setting a problem to yourself is the beginning of the solution, the essential first move in the game. It is a move in the nature of a decision.

(2) *Selective attention.* You need not tell me that you have set that problem to yourself, you need not tell it to yourself; your whole behavior will show that you did. Your mind becomes selective; it becomes more accessible to anything that appears to be connected with the problem, and less accessible to anything that seems unconnected. You eagerly seize upon any recollection, remark, suggestion, or fact that could help you to solve your problem, and you shut the door upon other things. When the door is so tightly shut that even the most urgent appeals of the external world fail to reach you, people say that you are absorbed.

(3) *Registering the pace of progress.* There is another thing that shows that you are seriously engaged in your problem; you become sensitive. You keenly feel the pace of your progress; you are elated when it is rapid, you are depressed when it is slow. Whatever comes to your mind is quickly

sized up: "It looks good," "It could help," or "No good," "No help." Such judgments are, of course, not infallible. (Although they seem to be more often correct than not, especially with talented or experienced people.) At any rate, such judgments and feelings are important for you personally; they guide your effort.

(4) *Where plausible reasoning comes in.* Let us see somewhat more concretely a typical situation.

You try to attain the solution in a certain direction, along a certain line. (For example, in trying to solve the geometrical problem of sect. 2 you reject fig. 16.2 and attempt to work with the more hopeful fig. 16.3.) You may feel quite keenly that you work in the right direction, that you follow a promising line of approach, that you are on the scent. You may feel so, by the way, without formulating your feeling in words. Or even if you say something such as, "It looks good," you do not take the trouble to analyze your confidence, you do not ask, "Why does it look good?" You are just too busy following up the scent.

Yet you may have bad luck. You run into difficulties, you do not make much progress, nothing new occurs to you and then you start doubting: "Was it a good start? Is this the right direction?" And then you may begin to analyze your feeling: "The direction looked quite plausible—but why is it plausible?" Then you may start debating with yourself, and some more distinct reasons may occur to you:

"The situation is not so bad. I could bring in a triangle. People always bring in triangles in such problems."

"It was probably the right start, after all. It looks like the right solution. What do I need for a solution with this kind of problem? Such a point—and I have it. And that kind of point—I have it too. And . . . "

It would be interesting to see more distinctly how people are reasoning in such a situation—in fact, it is our main purpose to see just that. Yet we need at least one more example to broaden our observational basis.

4. Deus ex machina.[2] The next example has to be a little less simple than that of sect. 2. It will be given in sect. 6 after some preparations in this section and the next. Sect. 6 will bring a proof, presented in a manner that contrasts with the usual manner of presentation. To emphasize the contrast, let us first see the proof as it would be presented in a (more advanced) textbook or in a mathematical periodical.

A mathematical book or lecture should be, first of all, correct and unambiguous. Still, we know from painful experience that a perfectly unambiguous and correct exposition can be far from satisfactory and may appear uninspiring, tiresome, or disappointing, even if the subject-matter presented is interesting in itself. The most conspicuous blemish of an

[2] Sect. 4, 5, and 6 reproduce, except for slight changes, parts of my paper "With, or without, motivation?" *American Mathematical Monthly*, v. 56, 1949, p. 684–691.

otherwise acceptable presentation is the "deus ex machina." Before further comments, I wish to give a concrete example. Let us look at the proof of the following not quite elementary theorem.[3]

If the terms of the sequence $a_1, a_2, a_3, \ldots$ are non-negative real numbers, not all equal to 0, then

$$\sum_1^\infty (a_1 a_2 a_3 \ldots a_n)^{1/n} < e \sum_1^\infty a_n.$$

Proof. Define the numbers $c_1, c_2, c_3, \ldots$ by

$$c_1 c_2 c_3 \ldots c_n = (n+1)^n$$

for $n = 1, 2, 3, \ldots$. We use this definition, then the inequality between the arithmetic and the geometric means (sect. 8.6), and finally the fact that the sequence defining e, the general term of which is $[(k+1)/k]^k$, is increasing. We obtain

$$\begin{aligned}
\sum_1^\infty (a_1 a_2 \ldots a_n)^{1/n} &= \sum_1^\infty \frac{(a_1 c_1 a_2 c_2 \ldots a_n c_n)^{1/n}}{n+1} \\
&\leqq \sum_1^\infty \frac{a_1 c_1 + a_2 c_2 + \ldots + a_n c_n}{n(n+1)} \\
&= \sum_{k=1}^\infty a_k c_k \sum_{n=k}^\infty \frac{1}{n(n+1)} \\
&= \sum_{k=1}^\infty a_k c_k \sum_{n=k}^\infty \left(\frac{1}{n} - \frac{1}{n+1}\right) \\
&= \sum_{k=1}^\infty a_k \frac{(k+1)^k}{k^{k-1}} \frac{1}{k} \\
&< e \sum_{k=1}^\infty a_k.
\end{aligned} \tag{d}$$

5. Heuristic justification. The crucial point of the derivation (d) is the definition of the sequence $c_1, c_2, c_3, \ldots$. This point appears right at the beginning without any preparation, as a typical "deus ex machina." What is the objection to it?

"It appears as a rabbit pulled out of a hat."

"It pops up from nowhere. It looks so arbitrary. It has no visible motive or purpose."

---

[3] I may be excused if I choose an example from my own work. See G. Pólya, Proof of an inequality, *Proceedings of the London Mathematical Society* (2), v. 24, 1925, p. LVII. The theorem proved is due to T. Carleman.

"I hate to walk in the dark. I hate to take a step when I cannot see any reason why it should bring me nearer to the goal."

"Perhaps the author knows the purpose of this step, but I do not, and therefore I cannot follow him with confidence."

"Look here, I am not here just to admire you. I wish to learn how to do problems by myself. Yet I cannot see how it was humanly possible to hit upon your . . . definition. So what can I learn here? How could I find such a . . . definition by myself?"

"This step is not trivial. It seems crucial. If I could see that it has some chances of success, or see some plausible provisional justification for it, then I could also imagine how it was invented and, at any rate, I could follow the subsequent reasoning with more confidence and more understanding."

The first answers are not very explicit, the later ones are better, and the last is the best. It reveals that an intelligent reader or listener desires two things:

First, to see that the present step of the argument is correct.

Second, to see that the present step is appropriate.

A step of a mathematical argument is appropriate if it is essentially connected with the purpose, if it brings us nearer to the goal. It is not enough, however, that a step *is* appropriate: it should *appear so* to the reader. If the step is simple, just a trivial, routine step, the reader can easily imagine how it could be connected with the aim of the argument. If the order of presentation is very carefully planned, the context may suggest the connection of the step with the aim. If, however, the step is visibly important, but its connection with the aim is not visible at all, it appears as a "deus ex machina" and the intelligent reader or listener is understandably disappointed.

In our example, the definition of c_n appears as a "deus ex machina." Yet this step is certainly appropriate. In fact, the argument based on this definition proves the proposed theorem, and proves it rather quickly and clearly. The trouble is that the step in question, although vindicated in the end, does not appear as justified from the start.

Yet how could the author justify it from the start? The complete justification takes some time; it is supplied by the full proof. What is needed is, not a complete, but an *incomplete* justification, a *plausible provisional ground*, just a hint that the step has some chances of success, in short, some *heuristic justification*.

6. The story of another discovery. It is almost unnecessary to remind the reader that the best stories are not true. They must contain, however, some essential elements of truth, otherwise they would not be any good. The following is a somewhat "rationalized" presentation of the steps that led me to the proof given in sect. 4. That is, the heuristic justification of the successive steps is suitably emphasized.

The theorem proved in sect. 4 is surprising in itself. We would be less surprised if we knew how it was discovered. We are led to it naturally in trying to prove the following: *If the series with positive terms*

$$a_1 + a_2 + a_3 + \ldots + a_n + \ldots$$

is convergent, the series

$$a_1 + (a_1a_2)^{1/2} + (a_1a_2a_3)^{1/3} + \ldots + (a_1a_2a_3 \ldots a_n)^{1/n} + \ldots$$

is also convergent. I shall try to emphasize some motives which may help us to find the proof.

(1) *A suitable known theorem.* It is natural to begin with the usual questions. *What is the hypothesis?* We assume that the series Σa_n converges—that its partial sums remain bounded—that

$$a_1 + a_2 + \ldots + a_n \text{ is not large.}$$

What is the conclusion? We wish to prove that the series $\Sigma(a_1, a_2 \ldots a_n)^{1/n}$ converges—that

$$(a_1a_2 \ldots a_n)^{1/n} \text{ is small.}$$

Do you know a theorem that could be useful? What we need is some relation between the sum of n positive quantities and their geometric mean. *Have you seen something of this kind before?* If you ever have heard of the inequality between the arithmetic and the geometric means, it has a good chance to occur to you at this juncture:

$$\text{(ag)} \qquad (a_1a_2 \ldots a_n)^{1/n} \leqq \frac{a_1 + a_2 + \ldots + a_n}{n}.$$

This inequality shows that $(a_1a_2 \ldots a_n)^{1/n}$ is small when $a_1 + a_2 + \ldots + a_n$ is not large. It has so many contacts with our problem that we can hardly resist the temptation of applying it:

$$\text{(a)} \qquad \begin{aligned} \sum_{n=1}^{\infty} (a_1\, a_2 \ldots a_n)^{1/n} &\leqq \sum_{n=1}^{\infty} \frac{a_1 + a_2 + \ldots + a_n}{n} \\ &= \sum_{k=1}^{\infty} a_k \sum_{n=k}^{\infty} \frac{1}{n} \end{aligned}$$

—complete failure! The series $\Sigma 1/n$ is divergent, the last line of (a) is meaningless.

(2) *Learning from failure.* It is difficult to admit that our plan was wrong. We would like to believe that at least some part of it was right. The useful questions are: *What was wrong with our plan? Which part of it could we save?*

The series $a_1 + a_2 + \ldots + a_n + \ldots$ converges. Therefore, a_n is small when n is large. Yet the two sides of the inequality (ag) are different when $a_1, a_2, \ldots a_n$ are not all equal, and they may be very different when $a_1, a_2, \ldots a_n$ are very unequal. In our case, a_1 is much larger than a_n, and so there may be a considerable gap between the two sides of (ag). This is probably the reason that our application of (ag) turned out to be insufficient.

(3) *Modifying the approach.* The mistake was to apply the inequality (ag) to the quantities

$$a_1, a_2, a_3, \ldots a_n$$

which are too unequal. Why not apply it to some related quantities which have more chance to be equal? We could try

$$1a_1, 2a_2, 3a_3, \ldots na_n.$$

This may be the idea! We may introduce such increasing compensating factors as 1, 2, 3, . . . n. We should, however, not commit ourselves more than necessary, we should reserve ourselves some freedom of action. We should consider perhaps, more generally, the quantities

$$1^\lambda a_1, 2^\lambda a_2, 3^\lambda a_3, \ldots n^\lambda a_n.$$

We could leave λ *indeterminate* for the moment, and choose the most advantageous value later. This plan has so many good features that it seems ripe for action:

$$\text{(b)} \qquad \begin{aligned} \sum_1^\infty (a_1 a_2 \ldots a_n)^{1/n} &= \sum_1^\infty \frac{(a_1 1^\lambda \cdot a_2 2^\lambda \ldots a_n n^\lambda)^{1/n}}{(1\cdot 2 \ldots n)^{\lambda/n}} \\ &\leqq \sum_{n=1}^\infty \frac{a_1 1^\lambda + a_2 2^\lambda + \ldots + a_n n^\lambda}{n(n!)^{\lambda/n}} \\ &= \sum_{k=1}^\infty a_k k^\lambda \sum_{n=k}^\infty \frac{1}{n(n!)^{\lambda/n}}. \end{aligned}$$

We run into difficulties. We cannot evaluate the last sum. Even if we recall various relevant tricks, we are still obliged to work with "crude equations" (notation $\approx$, instead of $=$):

$$(n!)^{1/n} \approx ne^{-1},$$

$$\begin{aligned} \sum_{n=k}^\infty \frac{1}{n(n!)^{\lambda/n}} &\approx e^\lambda \sum_{n=k}^\infty n^{-1-\lambda} \\ &\approx e^\lambda \int_k^\infty x^{-1-\lambda} dx \\ &= e^\lambda \lambda^{-1} k^{-\lambda}. \end{aligned}$$

Introducing this into the last line of (b) we come very close to proving

$$\text{(b}'\text{)} \qquad \sum_1^\infty (a_1 a_2 \ldots a_n)^{1/n} \leqq C \sum_1^\infty a_k$$

where C is some constant, perhaps $e^\lambda \lambda^{-1}$. Such an inequality would, of course, prove the theorem in view.

(4) *Looking back* at the foregoing reasoning we are led to repeat the question: "Which value of λ is the most advantageous?" Probably the λ that makes $e^\lambda \lambda^{-1}$ a minimum. We can find this value by differential calculus:

$$\lambda = 1.$$

This suggests strongly that the most obvious choice is the most advantageous: the compensating factor multiplying a_n should be $n^1 = n$, or some quantity not very different from n when n is large. This may lead to the simple value $C = e$ in (b′).

(5) *More flexibility.* We left λ indeterminate in our foregoing reasoning (b). This gave our plan a certain *flexibility*: the value of λ remained at our disposal. Why not give our plan still more flexibility? We could leave the compensating factor that multiplies a_n quite indeterminate; we call it c_n, and we will dispose of its value later, when we shall see more clearly what we need. We embark upon this further modification of our original approach:

$$\text{(c)}\qquad \begin{aligned} \sum_1^\infty (a_1 a_2 \ldots a_n)^{1/n} &= \sum_{n=1}^\infty \frac{(a_1c_1 \cdot a_2c_2 \ldots a_nc_n)^{1/n}}{(c_1c_2 \ldots c_n)^{1/n}} \\ &\leqq \sum_{n=1}^\infty \frac{a_1c_1 + a_2c_2 + \ldots + a_nc_n}{n(c_1c_2 \ldots c_n)^{1/n}} \\ &= \sum_{k=1}^\infty a_kc_k \sum_{n=k}^\infty \frac{1}{n(c_1c_2 \ldots c_n)^{1/n}}. \end{aligned}$$

How should we choose c_n? This is the crucial question and we can no longer postpone the answer.

First, we see easily that a factor of proportionality must remain arbitrary. In fact, the sequence $cc_1, cc_2, \ldots, cc_n, \ldots$ leads to the same consequences as $c_1, c_2, \ldots, c_n, \ldots$.

Second, our foregoing work suggests that both c_n and $(c_1c_2 \ldots c_n)^{1/n}$ should be asymptotically proportional to n:

$$c_n \sim Kn, \quad (c_1c_2 \ldots c_n)^{1/n} \sim e^{-1} Kn = K'n.$$

Third, it is most desirable that we should be able to effect the summation

$$\sum_{n=k}^{n=\infty} \frac{1}{n(c_1c_2 \ldots c_n)^{1/n}}.$$

At this point, we need whatever previous knowledge we have about simple series. If we are familiar with the series

$$\sum \frac{1}{n(n+1)} = \sum \left(\frac{1}{n} - \frac{1}{n+1}\right)$$

it has a good chance to occur to us at this juncture. This series has the property that its sum has a simple expression not only from $n = 1$ to $n = \infty$, but also from $n = k$ to $n = \infty$—a great advantage! This series suggests the choice

$$(c_1 c_2 \dots c_n)^{1/n} = n + 1.$$

Now, visibly $n + 1 \sim n$ for large n—a good sign! What about c_n itself? As

$$c_1 c_2 \dots c_{n-1} c_n = (n+1)^n, \qquad c_1 c_2 \dots c_{n-1} = n^{n-1},$$

$$c_n = \frac{(n+1)^n}{n^{n-1}} = \left(1 + \frac{1}{n}\right)^n n \sim en;$$

the asymptotic proportionality with n is a good sign. And the number e arises—a very good sign!

We choose this c_n and, after this choice, we take up again the derivation (d) in sect. 4 with more confidence than before.

Now, we may understand how it was humanly possible to discover that definition of c_n which appeared in sect. 4 as a "deus ex machina." The derivation (d) became also more understandable. It appears now as the last, and the only successful, attempt in a chain of consecutive trials, (a), (b), (c), and (d). And the origin of the theorem itself is elucidated. We see now how it was possible to discover the rôle of the number e which appeared so surprising at the outset.

7. Some typical indications. We have seen two examples in the foregoing. We examined first a "problem to find" (in sect. 2) then a "problem to prove" (in sect. 6.)[4] A much greater variety of examples were needed to illustrate properly the rôle of plausible reasoning in devising a plan of the solution. At any rate, from our examples we can disentangle a few typical circumstances indicative of the worth of a plan. In dealing with other circumstances of this kind we shall appeal to whatever experience the reader had in solving mathematical problems.

In enumerating such indicative circumstances, we shall not attempt completeness. In some cases we shall find it necessary to distinguish between problems to find and problems to prove. In such cases we shall give two parallel formulations, and give the formulation relative to problems to find first.

We consider a situation in which plausible reasoning comes naturally to the problem-solver. You are engaged in an exciting problem. You have conceived a plan of the solution, but somehow you do not like it quite. You have your doubts, you are not quite convinced that your plan is workable. In debating this matter with yourself, you are, in fact, examining a conjecture:

A. This plan of the solution will work.

Several pros and cons may occur to you as you examine your plan from various angles. Here are some conspicuous typical indications that may speak for the conjecture A.

---

[4] For this terminology, see *How to Solve It*, p. 141–144.

B_1. *This plan takes all the data into account.*

This formulation applies to problems to find. There is a parallel formulation that applies to problems to prove: *this plan takes into account all the parts of the hypothesis.* For example, fig. 16.3 combines all the data, and this is a good sign. Also fig. 16.2 contains all the data, but there is a difference between the two figures that the next points may clarify.

B_2. *This plan provides for a connection between the data and the unknown.*

There is a parallel formulation concerned with problems to prove: *this plan provides for a connection between the hypothesis and the conclusion.* For example, in sect. 6 (1), the inequality between the arithmetic and geometric means promised to create a connection between the hypothesis and the conclusion, and this promise moved us to work with that inequality. Fig. 16.3 appears to provide for a closer connection, and so it appears more hopeful, than fig. 16.2.

B_3. *This plan has features that are often useful in solving problems of this kind.*

For example, the plan starting from fig. 16.3 introduces at a more mature stage (fig. 16.4) the construction of a triangle. This is a good sign, since problems of geometric construction are often reduced to the construction of triangles.

B_4. *This plan is similar to one that succeeded in solving an analogous problem.*

B_5. *This plan succeeded in solving a particular case of the problem.*

For example, you have a plan to solve a difficult problem concerned with an arbitrary closed curve. Carrying through the plan seems to involve a lot of work and so you hesitate. Yet you observe that in the particular case when the closed curve is a circle your plan works and yields the right result. This is a good sign, and you feel encouraged.

B_6. *This plan succeeded in solving a part of the problem* (in finding some of the unknowns, or in proving a weaker conclusion).

This list is by no means exhaustive. There are still other typical indications and signs, but we need not list them here. At any rate, it would be useless to list them without proper illustration.[5]

8. Induction in invention. The problem-solver's conjecture A (that his plan of the solution will work) may be supported by one, or two, or more indications B_1, B_2, B_3, . . . (of the kind listed in the foregoing sect. 7). Such indications may occur to the problem-solver successively, each indication intensifying his confidence in his plan. Our foregoing discussions lead us to compare such a problem-solving process with the inductive process in which an investigator, examining a conjecture A, succeeds in verifying several consequences B_1, B_2, B_3, . . . in succession. We may also compare it with the legal procedure in which the jury examining an accusation A, takes

[5] Cf. *How to Solve It*, p. 212–221.

cognizance of several corroborating circumstances B_1, B_2, B_3, . . . in succession. We should not naïvely expect the identity of the three processes, but we should examine their similarity or dissimilarity with an open mind.

(1) When the problem-solver debates his plan of the solution with himself, this plan is usually more "fluid" than "rigid," it is more felt than formulated. In fact it would be foolish of the problem-solver to fix his plan prematurely. A wise problem-solver does not commit himself to a rigid plan. Even at a later stage, when the plan is riper, he keeps it ready for modification, he leaves it a certain flexibility, he reckons with unforeseen difficulties to which he might be obliged to adapt his plan. Therefore, when the problem-solver investigates the workability of his plan, he examines a changeable, sometimes a fleeting, object.

On the other hand, the conjectures that the mathematician or the naturalist investigates are usually pretty determinate: they are clearly formulated, or at least reasonably close to a clear formulation. Also the jury has a pretty determinate conjecture to examine: an indictment, the terms of which have been carefully laid down by the prosecution.

Let us note this striking difference that separates the problem-solver's investigation of the workability of his plan from the inductive investigation of a mathematical or physical conjecture, or from the judicial investigation of a charge: it is the difference between a changeable, or fleeting, and a determinate, relatively well defined object.

(2) The proceedings and acts of a court of justice are laid down in the record. The conjecture examined by the naturalist, and the evidence gathered for or against it, are also destined for a permanent record. Not so the problem-solver's conjecture concerning the workability of his scheme, or the signs speaking for or against it: their importance is ephemeral. They are extremely important as long as they guide the problem-solver's decisions. Yet, when the problem-solver's work enters a new phase, the plan itself may change, and then the indications speaking for or against it lose almost all interest. At the end, when the solution is attained and the problem is done, all such accessories are cast away. The final form of the solution may be recorded, yet the changing plans and the arguments for or against them are mostly or entirely forgotten. The building erected remains in view, but the scaffoldings, which were necessary to erect it, are removed.

Let us note this aspect of the difference between an inductive, or judicial, investigation on the one hand, and the problem-solver's appraisal of the prospects of his plan on the other: one is, the other is not, for permanent record.

(3) The conjecture A and the indications B_1, B_2, . . . listed in sect. 7 can be interpreted with a certain latitude. After the foregoing remarks (under (1) and (2)) we should not expect that a sharply defined interpretation will be too often applicable. Still, there is some advantage in beginning with

such an interpretation. We consider the problem-solver's conjecture A and an indication B supporting it, stated as follows:

A. This plan of the solution will work in its present form.

B. This plan of the solution takes into account all the data.

In order to describe the situation more precisely, we add: *It is known that each of the data is necessary.* If this is so

$$A \text{ implies } B.$$

In fact, if the plan should work and yield the correct solution, it must use all the data, each of which is necessary to the solution.

Now it is important to visualize clearly the situation: A is a conjecture in which the problem-solver is naturally interested, B is a statement that may, or may not, be true. Let us examine both possibilities.

(4) If all the data are necessary to the solution, but our plan of the solution does not take into account all the data, our plan, in its present form, cannot work. (It could work in a modified form.) That is, if B is false, A must also be false.

Now it is important to observe that we could reach this conclusion also by formal reasoning. In fact, we have followed here a classical elementary pattern of reasoning (already quoted in sect. 12.1) the "modus tollens" of the so-called hypothetical syllogism:

$$\begin{array}{c} A \text{ implies } B \\ B \text{ false} \\ \hline A \text{ false} \end{array}$$

(5) If, however, our plan of the solution does take into account all the data, it is natural to regard this circumstance as a good sign, as a favorable omen, as a forecast that our plan might work. (I imagine the problem-solver's relief when he notices that a datum that at first seemed to be neglected by his plan is used by it after all, and used in a neat manner, too.) In short, if B is true, A becomes more credible.

Now it is important to observe that, in fact, we could have reached this conclusion by simply following our fundamental inductive pattern:

$$\begin{array}{c} A \text{ implies } B \\ B \text{ true} \\ \hline A \text{ more credible} \end{array}$$

(6) We consider now another situation. It is similar to, but clearly different from, the situation explained under (3) and discussed under (4) and (5). Again we are concerned with the problem-solver's conjecture A

and an indication B supporting it. Yet the situation is now different (less sharply defined). A and B have the meanings:

A. This plan of the solution will work (in a modified form, perhaps).

B. This plan of the solution takes into account all the data.

In order to characterize the situation more fully, we have to add: *We strongly suspect, although we do not definitely know, that all the data are necessary.*

As above, A is a conjecture in which the problem-solver has a stake, and B is a statement that may, or may not, be true. We have to examine both possibilities.

If B turns out to be false, there is an argument against A, but it is not quite decisive. As B is false, our plan does not take into account all the data; nevertheless, we may stick to our plan (if we have some strong, although unexpressed, ground for it). There may be some (not yet clarified) ground to hope that a modification of our plan will take care of all the data eventually. Also the doubt that all the data may not be necessary could have some little influence.

If B turns out to be true, we can take this circumstance for an encouraging sign. In fact, even if A does not imply B, and so B is not absolutely certain with A, it may be that still

B with A is readily credible,

B without A is less readily credible.

In such a case the verification of B still counts as a sort of circumstantial evidence for A. (Cf. sect. 13.13 (5).)

(7) In the foregoing subsections (3), (4), (5), and (6), we discussed the indication listed under B_1 in sect. 7. (We called it just B.) The discussion of the other indications listed in sect. 7 (under B_2, B_3, . . . B_6) would disclose a similar picture.

As we have seen, A may imply B_1, but even if this is not so and B_1 is not necessarily associated with A, the chances may be strong that A will be accompanied by B_1. The relation of A to B_2 (or B_3, or B_4, . . .) is of the same nature. If the problem-solver's plan is any good, it must create some connection between the data and the unknown (or between the hypothesis and the conclusion); cf. B_2. It is not absolutely necessary that the solution should be similar to the solution of some formerly solved similar problem, yet the chances are usually pretty strong that this should be so; cf. B_3, B_4. If the plan works for the whole problem it must work, of course, for any particular case or any portion of the problem; cf. B_5, B_6.

Therefore, if we are in doubt about A, but succeed in observing B_1 or B_2 or B_3 . . . , we can reasonably regard our observation as some kind of inductive or circumstantial evidence for A, as an indication in favor of the problem-solver's conjecture that his plan will work.

(8) If, in spite of much work, the naturalist succeeds in verifying only a few not too surprising consequences of his conjecture, he may be moved to drop it. If too little evidence is submitted against the defendant, the court may drop the case. If, after a long and strenuous effort, only a few weak indications in favor of his plan have occurred to the problem-solver, he may be moved to modify his plan radically, or even drop it altogether.

On the other hand, if several consequences are verified, several pieces of evidence against the defendant submitted, several indications observed, the case for the naturalist's conjecture, for the prosecution, or for the problem-solver's plan may be greatly strengthened. Yet even more important than the number may be the variety. Consequences that are very different from each other, witnesses who are obviously independent, indications that come from different sides, count more heavily. (Cf. sect. 12.2, 13.11, 15.9.)

(9) In spite of such similarities, there is a considerable difference. The naturalist's task is to gather as much evidence as he can for or against his conjecture. The court's task is to examine all the relevant evidence submitted. Yet it is not the problem-solver's task to collect as much evidence as he can for or against his plan of the solution, or to debate such evidence to the bitter end: his task is to solve the problem, by any means, in following this plan of the solution, or any other plan, or no plan.

Even a faulty plan may serve the problem-solver's purpose. To solve his problem, he has to mobilize and organize the relevant parts of his past experience. Working with a faulty plan, but with genuine effort, the problem-solver may stir up some relevant item which otherwise would have remained hidden and unawakened in the background; this may give him a new departure. In problem-solving a bad plan is frequently useful; it may lead to a better plan.

(10) Two persons presented with the same evidence may honestly disagree, even if they are relying on the same patterns of plausible reasoning. Their backgrounds may be different. My unexpressed, inarticulate grounds, my whole background may influence my evaluation of experimental or judicial evidence. They may influence still more my evaluation of the indications for or against my plan of the solution, and this is not unreasonable. It is reasonable that, working at the solution of a problem, I should attach more weight than under other conditions to the promptings of my background and less weight to distinctly formulated grounds: to stir up relevant material hidden somewhere in the background is the thing I am working for.

Still, it seems to me that one of the principal assets of a seasoned problem-solver is that he is able to deal astutely with indications for or against the workability of his plan, as a well-trained naturalist deals with experimental evidence, or an experienced lawyer with legal evidence.

9. A few words to the teacher. Mathematics has many aspects. To many students, I am afraid, mathematics appears as a set of rigid rules, some of which you should learn by heart before the final examinations, and all of which you may forget afterwards. To some instructors, mathematics appears as a system of rigorous proofs which, however, you should refrain from presenting in class, but instead present some more popular although inconclusive talk of which you are somewhat ashamed. To a mathematician, who is active in research, mathematics may appear sometimes as a guessing game: you have to guess a mathematical theorem before you prove it, you have to guess the idea of the proof before you carry through the details.

To a philosopher with a somewhat open mind all intelligent acquisition of knowledge should appear sometimes as a guessing game, I think. In science as in everyday life, when faced by a new situation, we start out with some guess. Our first guess may fall wide of the mark, but we try it and, according to the degree of success, we modify it more or less. Eventually, after several trials and several modifications, pushed by observations and led by analogy, we may arrive at a more satisfactory guess. The layman does not find it surprising that the naturalist works in this way. The knowledge of the naturalist may be better ordered with a view to selecting the appropriate analogies, his observations may be more purposeful and more careful, he may give more fancy names to his guesses and call them "tentative generalizations," but the naturalist adapts his mind to a new situation by guessing like the common man. And the layman is not surprised to hear that the naturalist is guessing like himself. It may appear a little more surprising to the layman that the mathematician is also guessing. The result of the mathematician's creative work is demonstrative reasoning, a proof, but the proof is discovered by plausible reasoning, by guessing.

If this is so, and I believe that this is so, there should be a place for guessing in the teaching of mathematics. Instruction should prepare for, or at least give a little taste of, invention. At all events, the instruction should not suppress the germs of invention in the student. A student who is somewhat interested in the problem discussed in class *expects* a certain kind of solution. If the student is intelligent, he foresees the solution to some extent: the result may look thus and so, and there is a chance that it may be obtained by such and such a procedure. The teacher should try to realize what the students might expect, he should find out what they do expect, he should point out what they should reasonably expect. If the student is less intelligent and especially if he is bored, he is likely to produce wild and irresponsible guesses. The teacher should show that guesses in the mathematical domain may be reasonable, respectable, responsible. I address myself to teachers of mathematics of all grades and say: *Let us teach guessing!*

I did not say that we should neglect proving. On the contrary, we should teach both proving and guessing, both kinds of reasoning, demonstrative and plausible. More valuable than any particular mathematical fact or trick, theorem, or technique, is for the student to learn two things:

First, to distinguish a valid demonstration from an invalid attempt, a proof from a guess.

Second, to distinguish a more reasonable guess from a less reasonable guess.

There are special cases in which it is more important to teach guessing than proving. Take the teaching of calculus to engineering students. (I have a long and varied experience of this kind of teaching.) Engineers need mathematics, quite a few of them have a healthy interest in mathematics, but they are not trained to understand ε-proofs, they have no time for ε-proofs, they are not interested in ε-proofs. To teach them the rules of calculus as a dogma imposed from above would not be educational. To pretend that your proof is complete when, in fact, it is not, would not be honest. Confess calmly that your proofs are incomplete, but give respectable plausible grounds for the incompletely proved results, from examples and analogy. Then you need not be ashamed of fake proofs, and some students may remember your teaching after the examinations. On the basis of a long experience, I would say that talented students of engineering are usually more accessible to, and more grateful for, well-presented plausible grounds than for strict proofs.

I said that it is desirable to teach guessing, but not that it is easy to teach it. There is no foolproof method for guessing, and therefore there cannot be any foolproof method to teach guessing. I may have said a few foolish things in the foregoing, perhaps, but I have avoided the most foolish thing, I hope, which would be to pretend that I have an infallible method to teach guessing.

Still, it is not impossible to teach guessing. I hope that some of the examples explained at length and some of the exercises proposed in the foregoing will serve as useful suggestions. These suggestions have the best chance to fall on fertile ground with teachers who have some real experience in problem-solving.

Take, for instance, the example treated in sect. 4 and 6. The two presentations, in sect. 4 and sect. 6, are very different. The most obvious difference is that one is short and the other long. The most essential difference is that one gives proofs and the other plausibilities. One is designed to check the *demonstrative conclusions* justifying the successive steps. The other is arranged to give some insight into the *heuristic motives* of certain steps. The demonstrative presentation follows the accepted manner, usual since Euclid; the heuristic presentation is extremely unusual in print. Yet an alert teacher can use both manners of exposition. In fact, he could construct, if need be, a third presentation that is between the two with due regard to the available time, to the interest of his students, to all the conditions under which he works.[6]

---

[6] For a presentation intermediate between sect. 4 and sect. 6 see G. H. Hardy, J. E. Littlewood, and G. Pólya, *Inequalities*, p. 249–250.

This book is principally addressed to students desiring to develop their own ability, and to readers curious to learn about plausible reasoning and its not so banal relations to mathematics. The interests of the teacher are not neglected, I hope, but they are met rather indirectly than directly. I hope to fill that gap some day. In the meantime, I reiterate my hope that this book, as it is, may be useful to some teachers, at least to those teachers who have some genuine experience in problem-solving. The trouble is that there are so few teachers of mathematics who have such experience. And even the best School of Education has not yet succeeded in producing the marvellous teacher who has such an excellent training in teaching methods that he can make his students understand even those things that he does not understand himself.

EXAMPLES AND COMMENTS ON CHAPTER XVI

1. *To the teacher: some types of problems.* This book is designed to serve various categories of readers: those who wish to understand guessing, those who wish to learn guessing, and those who wish to teach guessing. The reader of the last category is seldom directly addressed, but an alert teacher could learn something both from the examples presented in this book and from the manner of presentation. He could see, for instance, that there are manners of proposing a problem very different from the usual manner. I wish to point out a few types of problems, the "guess-and-prove" problems, the "test-consequences" problems, the "you-may-guess-wrong" problems, and the "small-scale-theory" problems. All these types could be used by an alert teacher to challenge his more intelligent students and to relieve the monotony of routine problems which fill the textbooks.

Guess and prove. Mathematical facts are first guessed and then proved, and almost every passage in this book endeavors to show that such is the normal procedure. If the learning of mathematics has anything to do with the discovery of mathematics, the student must be given some opportunity to do problems in which he first guesses and then proves some mathematical fact on an appropriate level. Still, the usual textbooks do not offer such an opportunity: ex. 1.2, 5.1, 5.2, 7.1–7.6 (and many others) do.

Test consequences. Philosophers and non-philosophers disagree about almost everything touching induction, but there is little doubt that the most usual inductive procedure consists in examining a general statement by testing its particular consequences. This inductive procedure is of daily use in mathematical research, and could be of daily use in the classroom with real benefit to the students. See sect. 12.2 and ex. 12.3–12.6. Cf. ex. 6.

You may guess wrong. You should acquire some experience in guessing. You should know from personal contact with the real thing that guesses may be respectable, that guesses may go wrong, and that even your own quite respectable guesses may go wrong. For such experience, solve ex. 11.1–11.12.

Small scale theory. On almost every page of this book some relatively elementary problem is discussed so that the discussion should shed some light upon a point that may arise in connection with other, not so elementary, problems. There is a reason to prefer such "small scale research": a less elementary problem could show the point in question on a more impressive scale, but it would demand a much longer explanation and much more preliminary knowledge. It is not too easy to "reduce the scale": elementary problems that show clearly enough the relevant features of plausible, or inventive, reasoning may be hard to find. It is also possible, but still less easy, to devise elementary problems to illustrate the scientist's activity in constructing a theory. The following ex. 2, 3, and 4 offer such "small-scale-theory problems"; ex. 5 and 6 are somewhat similar.

2. A quadrilateral is cut into four triangles by its two diagonals. We call two of these triangles "opposite" if they have a common vertex but no common side. Prove the following statements:

(a) The product of the areas of two opposite triangles is equal to the product of the areas of the other two opposite triangles.

(b) The quadrilateral is a trapezoid if, and only if, there are two opposite triangles equal in area.

(c) The quadrilateral is a parallelogram if, and only if, all four triangles are equal in area.

3. (a) Prove the following theorem: A point lies inside an equilateral triangle and has the distances x, y, and z from the three sides, respectively; h is the altitude of the triangle. Then $x + y + z = h$.

(b) State precisely and prove the analogous theorem in solid geometry concerning the distances of an inner point from the four faces of a regular tetrahedron.

(c) Generalize both theorems so that they should apply to any point in the plane or space, respectively (and not only to points inside the triangle or tetrahedron). Give precise statements, and also proofs.

4. Consider the following four propositions, (I)–(IV), which are not necessarily true.

(I) If a polygon inscribed in a circle is equilateral it is also equiangular.

(II) If a polygon inscribed in a circle is equiangular it is also equilateral.

(III) If a polygon circumscribed about a circle is equilateral it is also equiangular.

(IV) If a polygon circumscribed about a circle is equiangular it is also equilateral.

(a) State which of the four propositions are true and which are false, giving a proof of your statement in each case.

(b) If, instead of general polygons, we should consider only quadrilaterals which of the four propositions are true and which are false?

(c) What about pentagons?

(d) Could you guess, or perhaps even prove, more comprehensive statements? They should explain your observations (b) and (c).

5. Let α, β, and γ denote the angles of a triangle. Show that

(a) $\sin \alpha + \sin \beta + \sin \gamma = 4 \cos \frac{\alpha}{2} \cos \frac{\beta}{2} \cos \frac{\gamma}{2}$,

(b) $\sin 2\alpha + \sin 2\beta + \sin 2\gamma = 4 \sin \alpha \sin \beta \sin \gamma$,

(c) $\sin 4\alpha + \sin 4\beta + \sin 4\gamma = - 4 \sin 2\alpha \sin 2\beta \sin 2\gamma$.

6. Consider the frustum of a right pyramid with square base. Call "midsection" the intersection of the frustum with a plane parallel to the base and the top and at the same distance from both. Call "intermediate rectangle" the rectangle of which one side is equal to a side of the base and the other side is equal to a side of the top.

Four different friends of yours agree that the volume of the frustum equals the altitude multiplied by a certain area, but they disagree and make four different proposals regarding this area:

(I) the midsection

(II) the average of the base and the top

(III) the average of the base, the top, and the midsection

(IV) the average of the base, the top, and the intermediate rectangle.

Let h be the altitude of the frustum, a the side of its base, and b the side of its top. Express each of the four proposed rules in mathematical notation, decide whether it is right or wrong, and prove your answer.

7. *Qui nimium probat, nihil probat.* That is, if you prove too much, you prove nothing. I cannot tell in which sense the inventor of this classical saying intended to use it, but I wish to explain a meaning of the sentence that I found extremely helpful when I started doing some work in mathematics, and ever since. The sentence reminds me of one of the most useful signs by which we can judge the workability of a plan of the solution.

Here is the situation: you wish to prove a proposition. This proposition consists of a conclusion and a hypothesis which has several clauses, and you know that each one of these clauses is necessary to the conclusion, that is, none of them can be discarded without rendering the conclusion of the proposition invalid. You have conceived a plan of the proof, and you are weighing the chances of your plan. If your plan does not bring into play all the clauses, you have to modify your plan or reject it: if it would work as it is and prove the conclusion, although it leaves aside this or that clause of the hypothesis, it would prove too much, that is, something false, and so it would prove nothing.

I said that your plan should bring into play those clauses. I mean that mere lip-service is not enough, just mentioning them does not do: your plan should provide for essential use of each clause in the proof. The framework intended to support the conclusion cannot stand up unless it has a solid foothold in each clause of the hypothesis.

It may be very difficult to devise a plan that duly brings into play all the clauses of the hypothesis. Therefore, if a plan promises to catch all those clauses, we greet it with relief: there is an excellent sign, a strong indication that the plan may work.

For corresponding remarks on the solution of "problems to find" all the data of which are necessary see sect. 8 (3), (4), (5).

If you prefer a French sentence to a Latin saying, here is one: "La mariée est trop belle"; the bride looks too good. I do not think that I need to amplify this; after the foregoing the reader can picture all the details to himself.

8. *Proximity and credibility.* How far away is the solution? How much remains to be done? Such questions lie heavily on the mind of the student who has to finish his task within a set time, but they are present in the mind of every problem-solver.

(1) We are even able to answer such questions to a certain extent, not precisely, of course, but rather correctly on the average, I am inclined to believe.

For instance, let us look back at figs. 16.1–16.5 and the process of solution that they represent. The problem-solver may feel that fig. 16.3 is much closer to the solution than fig. 16.2; and he may feel that the solution is within easy reach once he has arrived at fig. 16.4.

In judging the proximity of the solution, we may rely on inarticulate feelings, or on more distinct signs. Any sign that indicates that our plan of the solution might work may be interpreted also as a sign of progress toward the solution, and help us to estimate the distance that we still have to go.

(2) Let us consider the solution of a "problem to prove." The aim is to prove (or disprove) a certain theorem. The problem-solver may trust, or distrust, the theorem at which he works. Yet, if he is any good as a problem-solver, he should be prepared to revise his beliefs. And so the questions, "Is the theorem credible? How credible is it?" are ever present in his mind, although sometimes more, sometimes less, in the foreground. If anything new comes in sight, he has two questions: "Does it render the theorem more, or less, credible?" "Does it bring the solution nearer or not?" His attention may be, of course, so absorbed by the new fact that he finds no time to formulate either question in words. He may also give the answer to himself without words. Even if he says "Good sign" or "Bad sign," he will scarcely take the trouble to express at length what he means. Is it a sign of the *proximity* of the solution, or a sign of the *credibility* of the theorem? Yet, if he is a good problem-solver, he knows well enough the important difference between proximity and credibility, and this difference will show up in his work, in his handling of the problem.

(3) Recalling a name once known, but now forgotten, is a task that is simpler than but somewhat similar to a mathematical problem. We can often observe people trying to recall a name, and we could learn a few interesting things from such observations.

In a conversation, your friend wants to tell you a name (the name of a shop, of an acquaintance, or of an author, perhaps). He gets stuck and you hear him say: "Just a minute and I shall get it," or "Wait a little, it may take a short while till I recall it," or "How stupid of me, but I am not able to recall it now, although I am sure it will occur to me in a few hours, perhaps tomorrow morning." Obviously, your friend tries to judge the proximity of that name, he tries to measure a sort of "psychological distance." I would surmise that his predictions will turn out about right, that his evaluation of that "psychological distance" is roughly correct.

With a view to a possible comparison with problem-solving, it is also interesting to note that a person who is not able to recall a name fully may be able to recall it partially or, better expressed perhaps, he may be able to recall certain features of the name. You may hear your friend say: "The name is not Battenberg—he is not the husband of the Queen, after all—but it is a German name, of three syllables, very similar to Battenberg." And (I have observed such cases) your friend may be completely right in all these particulars, although the correct name will occur to him only a few days later.

Quite similarly, a mathematician, although he has not yet solved his problem, may foresee certain features of the solution quite reliably.

Those aspects of problem-solving that are the most interesting for the future mathematician or the teacher are not readily accessible to the usual methods of experimental psychology. Perhaps, recalling a name, a process in some respects similar to solving a mathematical problem, could be brought more easily within the scope of psychological experiments.

9. *Numerical computation and plausible reasoning.* Although numbers are often regarded as symbols of the highest attainable certainty, the results of numerical computation are by no means certain; they are only plausible. Numerical computation depends on plausible reasoning in many ways.

(1) You have to do a long numerical computation. The final result is attained in a sequence of steps. You have a very good chance to do correctly any single step, yet there are many steps, there is a possibility of a mistake at each step, and the final result may be wrong. How can you protect yourself from error?

Compute the desired number twice by procedures as different as possible. If the two computations yield different results, it is certain that at least one of them is wrong, but both may be wrong. If the two computations agree, it is by no means certain that the result twice obtained is correct, but it may be correct, and the agreement is an indication of its correctness. The weight of this indication depends on the difference between the two procedures used.

For example, the weight is very small if, having done the computation once, you repeat it immediately afterwards, without any change in the method: with the first computation still fresh in your mind and in your fingers, you can easily stumble a second time at the same place where you stumbled the first time. To repeat the computation after a while is a little better, to let another person do it the second time is considerably better, to do the second computation by a very different method is still better.

In fact, if two quite different procedures arrive at the same result, we have only two obvious conjectures: the result may be correct, or the agreement may be due to chance. If the probability of an agreement by mere chance is very small, the second of the two rival conjectures is correspondingly unlikely, we are inclined to reject it, and so we are disposed to place more confidence in the first conjecture, that is, we may rely more on the correctness of the result.

The more the procedures of the two computations differ, the more realistic is the simplest evaluation of the probability of their agreement: the probability is 10^{-n} that the two computations arrive by mere chance at the same n figures; cf. sect. 14.9 (3), ex. 14.11, but also ex. 14.32, and ex. 12.

In large-scale computation it is good practice to render a chance coincidence still more improbable by the introduction of *multiple controls.* Two computations are performed by methods as different as conveniently possible, yet so that they should agree, if correct, not only in the final result, but also in several intermediate results. As the agreement increases, it becomes increasingly difficult to attribute it to mere chance although, of course, chance can never be fully excluded and the result can never be fully guaranteed.

(2) In the foregoing we tacitly assumed that the two computations of which we compare the results are known to be strictly equivalent theoretically. Yet in applied mathematics we often have to work with approximations, and we may compare numerical results which need not agree completely even if all arithmetical operations involved are faultless; we just hope that they will agree "roughly." Moreover, the theory of the method of approximation with which we work may be very imperfectly known. Under such circumstances, of course, the scope of plausible reasoning is even wider and such reasoning is more hazardous. Cf. ex. 11.23.

(3) Two mathematicians, A and B, investigated the same set of nine combinatorial problems. We need not know the subject of these problems (they deal with the hypercube in four dimensions) but it is important to know that they are arranged in order of increasing difficulty. The first two problems are trivial, the third problem is easy, the fourth is less easy, then they become increasingly more complex, and the last problem is the hardest.

Both A and B solved the problems, but their results did not quite agree. Here are their solutions for the nine problems:

$$A: \; 1 \quad 1 \quad 4 \quad 6 \quad 19 \quad 27 \quad 47 \quad 55 \quad 78$$
$$B: \; 1 \quad 1 \quad 4 \quad 6 \quad 19 \quad 27 \quad 50 \quad 56 \quad 74.$$

That is, A and B agree about the first six problems, which are easier, but disagree concerning the last three problems, which are harder. In fact, they followed very different methods.

A attacked each of the nine problems independently of the others. His method for each problem is somewhat different, and as he proceeds to harder problems, also his method becomes increasingly complex.

B attacked the problems by a uniform method. His work consists of two parts. The first part, which is more difficult, is a common preparation for the solution of all nine problems. The second part, which is more routine, applies the result of the first part to each single problem according to a uniform rule. Treated with B's method, the problems appear to differ in difficulty much less than with A's method.

It seems to me that the situation described gives us a reasonable ground to trust B's solution more than A's solution.

Since the two methods, which are very different, agree with regard to the first six problems, and these problems are easier anyhow, there is good ground to believe that the solution of these problems is correct. About the solution of the first three problems there is no doubt.

Since the result of the first part of B's work is verified by its consequences in 3 cases out of 9, and is presumably verified in 3 more cases (it is neither verified nor refuted in the 3 remaining cases) there is good ground to trust this result.

If, however, the first part of B's work were correct (as it presumably is) he could only err in the second, more routine, part in dealing with the last three problems. Yet A had the greatest difficulty in dealing with them. And so A appears to have more chances of error than B.[7]

The case just discussed is rather special, but it shows that there are further possibilities in exploring the patterns of plausible reasoning. For example, it may be a rewarding task to express the plausible argument just presented in formulas of the Calculus of Probability as fittingly as possible.

10. You have to add a column of ten six-place numbers, beginning so, for instance:

1 5 9 6 . 0 3
1 6 4 6 . 0 7
1 7 8 1 . 1 0
. . . .

Describe various procedures to do it.

[7] See G. Pólya, Sur les types des propositions composées, *Journal of Symbolic Logic*, vol. 5 (1940), p. 98–103.

11. Call "elementary step" the addition of a one-digit number that you see written to a two-digit number that you have in mind; include, however, the possibility (which renders the step easier) that the second number is also written, or that it has one digit only. How many elementary steps are required to perform the addition mentioned in ex. 10 in the most usual manner?

12. In doing a computation, you obtain first two nine-place numbers, and then the final result as their difference, which turns out to be a three-place number. Another procedure of computation obtains the same three-place result as the difference of two seven-place numbers. Compute the probability that such an agreement is due to chance, using the formula given in ex. 9 (1).

13. *Formal demonstration and plausible reasoning.* You have to do a long numerical computation. The final result is attained in a sequence of steps and must be correct if each step is correct. Each single step (as the addition $3 + 7$ or the multiplication 3×7) is so simple and familiar that you cannot slip under somewhat favorable circumstances, when your attention is "undivided." Still, like everyone else, you are liable to make mistakes in a computation. After having performed the successive steps quite carefully, you should not trust the final result without checking it.

You go through a lengthy mathematical demonstration. The demonstration is supposed to be decomposed into steps each of which you can check perfectly, and the final conclusion must be correct if each step is correct. Yet you may make mistakes like everybody else. After having checked the successive steps quite carefully, can you trust the final conclusion? Not more, and perhaps less, than the final result of a long computation.

In fact, a mathematician who has checked the details of a demonstration step by step and has found each step in order may be still dissatisfied. He needs something more to satisfy himself than the correctness of each detail. What?

He wants to *understand* the demonstration. After having struggled through the proof step by step, he takes still more trouble: he reviews, reworks, reformulates, and rearranges the steps till he succeeds in grouping the details into an understandable whole. Only then does he start trusting the proof.

I would not dare to analyze what constitutes "understanding." Some people say that it is based on "intuition" and they credit intuition with perceiving the whole and grouping the details into a well-arranged harmonious whole. I would not dare to contradict this, although I have some misgivings.[8] Yet I wish to call attention to a point strongly suggested by the examples and discussions in this book.

[8] The meaning of intuition and its rôle in grouping the details is usually not too well explained. It is remarkable, however, that Descartes, with whom the modern usage of the term "intuition" originates, explains both points rather impressively, in the Third and the Seventh of his Rules for the Direction of the Mind. See *Oeuvres*, edited by Adam and Tannery, vol. 10, p. 368–370 and 387–388.

Some practice may convince us that analogy and particular cases can be helpful both in finding and in understanding mathematical demonstrations. The general plan, or considerable parts, of a proof may be suggested or clarified by analogy. Particular cases may suggest a proof (see, for instance, sect. 3.17); on the other hand, we may test an already formulated proof by observing how it works in a familiar or critical particular case. Yet analogy and particular cases are the most abundant sources of plausible argument: perhaps, they not only help to shape the demonstrative argument and to render it more understandable, but also add to our *confidence* in it. And so we are led to suspect that a good part of *our reliance on demonstrative reasoning may come from plausible reasoning.*

SOLUTIONS

Solutions

SOLUTIONS, CHAPTER XII

1. The resemblance to the table of sect. 3.12 (partitions of space) can be regarded as closer. All the tables mentioned, except that in sect. 3.1 that lists polyhedra, are concerned with inductive evidence supporting a proposition A of a particular nature: A asserts that a certain statement S_n, the meaning of which depends on a variable integer n, is true for $n=1, 2, 3, \ldots$. Inductive examination of such a proposition A naturally proceeds in a certain order: we test S_1 first, then S_2, then S_3, and so on. This order is visible in the arrangement of the tables. Yet if we examine a proposition concerned with polyhedra, as in sect. 3.1, there is no such "natural" order. We may begin our investigation with the tetrahedron which, from various viewpoints, may be regarded as the "simplest" polyhedron. Yet which polyhedron should we examine next? There is no convincing ground to regard some polyhedron as the "second in simplicity," another as the third, and so on.

2. If the cases 101 and 301 appear "further removed" from the already verified cases 1, 2, 3, . . . 20 than the cases 21 and 22 (and in Euler's investigation they do appear so), it seems reasonable (and it is certainly in harmony with the pattern introduced in sect. 2) to attach more weight to the verification of the cases 101 and 301 than to that of the cases 21 and 22.

3. (1) $c = p$: the triangle degenerates into a straight line; $A = 0$.

(2) $c > p$ makes A imaginary: there is no triangle with $c > p$.

(3) $a = b = c$: the triangle is equilateral and $A^2 = 3a^4/16$, which is correct.

(4) $a^2 = b^2 + c^2$: the triangle is a right triangle and

$$\begin{aligned} 16A^2 &= (a + b + c)\,(b + c - a)\,(a - b + c)\,(a + b - c) \\ &= [(b + c)^2 - a^2]\,[a^2 - (b - c)^2] \\ &= (2bc)^2 \end{aligned}$$

or $A^2 = b^2c^2/4$, which is correct.

(5) $b = c = (h^2 + a^2/4)^{1/2}$: the triangle is isosceles, with height h, and

$$16A^2 = (a + 2b)(2b - a)a^2$$
$$= (4b^2 - a^2)a^2$$
$$= 4h^2a^2,$$

which is correct.

(6) The dimension is correct.

(7) The expression for A^2 is symmetric in the three sides a, b, and c, as it should be.

4. (1) $d = 0$: the quadrilateral becomes a triangle, the asserted formula reduces to Heron's formula, ex. 3.

(2) $d = p$: the quadrilateral degenerates into a straight line; $A = 0$.

(3) $d > p$ makes A imaginary: there is no quadrilateral with $d > p$.

(4) $a = b = c = d$ yields $A^2 = a^4$, which is correct for a square, but incorrect (too large) for a rhombus: the square is inscribable in a circle, the rhombus is not.

(5) $c = a$, $d = b$ yields $A^2 = (ab)^2$ which is correct for a rectangle but incorrect (too large) for an oblique parallelogram: the rectangle is inscribable in a circle, the oblique parallelogram is not.

(6) In the foregoing cases (4) and (5) the asserted formula attributes a value too large to the area of non-inscribable quadrilaterals: this agrees with sect. 10.5 (2) and 10.6 (3).

(7) The asserted formula gives the dimension of A correctly.

(8) According to the asserted formula A is symmetric in the four sides a, b, c, and d: an inscribed quadrilateral remains inscribed in the same circle if two neighboring sides are interchanged. (Consider the four isosceles triangles with common vertex at the center of the circle of which the bases are the four sides.)

These remarks do not prove the proposed formula, of course, but they render it very plausible, according to the pattern exhibited in sect. 2. For a proof see ex. 8.41.

5. (1) $a = b = c = e = f = g$: the tetrahedron is regular; $V = 2^{1/2}a^3/12$.

(2) $e^2 = b^2 + c^2$, $f^2 = c^2 + a^2$, $g^2 = a^2 + b^2$: the tetrahedron is "tri-rectangle," that is, the three edges a, b, and c, starting from the same vertex, are perpendicular to each other; $V = abc/6$.

(3) $e = 0$, $b = c$, $f = g$: the tetrahedron collapses, becomes a plane figure, a triangle; $V = 0$.

(4) $e = a$, $f = b$, $g^2 = c^2 = a^2 + b^2$: the tetrahedron collapses, becomes a rectangle with sides a and b; $V = 0$.

(5) A particular case more extended than (4): the tetrahedron becomes a plane quadrilateral, with sides a, b, e, f and diagonals c, g. Then $V = 0$ yields a relation between the sides and the diagonals of a general quadrilateral which also can be verified directly, although less easily.

(6) The dimension is correct.

(7) The expression of V is *not* symmetric in all six edges, but need not be: the three edges a, b, c which start from the same vertex do not play the same rôle as the three edges e, f, g which include a triangle (a face). We can transform, however, the proposed expression into the following:

$$\begin{aligned} 144V^2 = {} & a^2e^2 (b^2 + f^2 + c^2 + g^2 - a^2 - e^2) \\ & + b^2f^2 (c^2 + g^2 + a^2 + e^2 - b^2 - f^2) \\ & + c^2g^2 (a^2 + e^2 + b^2 + f^2 - c^2 - g^2) \\ & - e^2f^2g^2 - e^2b^2c^2 - a^2f^2c^2 - a^2b^2g^2. \end{aligned}$$

The first three lines correspond to the three pairs of opposite edges, the four terms in the last line to the four faces of the tetrahedron: the new algebraic form exhibits all the symmetry (interchangeability) of the data due to the geometric configuration. By the way, the new form (the correctness of the algebraic manipulation that yielded it) can also be checked by the particular cases (1), (2), (3), (4).

6. (1) We find for

$$s_1, \quad s_3, \quad s_5, \quad q, \quad r$$

the numerical values

$$6, \quad 36, \quad 276, \quad 11, \quad 6$$

respectively, which verify both formulas.

(2) More generally, let

$$a + b + c = s_1 = p = 0$$

Then

$$\begin{aligned} s_3 &= -3ab(a + b), \\ s_5 &= -5ab(a + b)(a^2 + ab + b^2), \\ q &= -a^2 - ab - b^2, \\ r &= -ab(a + b) \end{aligned}$$

and both formulas are verified since

$$-\frac{3s_5}{5s_3} = -a^2 - ab - b^2, \qquad \frac{s_3}{3} = -ab(a + b).$$

(3) More generally, let $c = 0$. A longer straightforward computation verifies both formulas.

(4) More generally, let $b + c = 0$. Then $s_1 = a$, $s_3 = a^3$, $s_5 = a^5$: denominator *and* numerator vanish in the expressions proposed for q and r.

For proof and generalization see *Journal des math. pures et appliquées*, ser. 9, vol. 31 (1952) p. 37–47.

7. The statement B_4 (with $r = h = 0$) is contained as a particular case both in B_2 ($r = 0$) and in B_3 ($h = 0$). Hence, if either B_2 or B_3 is true, B_4 must be true too. If we have this clearly in mind, the verification of B_4, coming after that of B_2 or B_3, does not convey to us new information. And, where there is no new information, there can be no new evidence, I should think. Still, it is desirable to observe B_4; it rounds off the picture.

8. Closer attention to the derivation in No. 9–13 of Euler's memoir in sect. 6.2 shows that $C_1^*, C_2^*, \ldots C_{20}^*$ follow from $C_1, C_2, \ldots C_{20}$ mathematically: therefore, the verification of $C_1^*, C_2^*, \ldots C_{20}^*$ did not really yield new information or evidence, but that of C_{101}^* and C_{301}^* did.

9. The pattern agrees essentially with the pattern that will be introduced in ex. 11; see comments there.

10. If $8n + 3 = w^2 + 2p$, the integer w is necessarily odd. Therefore, w^2 is of the form $8n + 1$ and so p of the form $4n + 1$. Euler proved that

$$p = u^2 + v^2;$$

of the two integers u and v, one must be odd and the other even. Hence

$$2p = 2u^2 + 2v^2 = (u + v)^2 + (u - v)^2.$$

Now, w, $u + v$, and $u - v$ are odd. Let

$$w = 2x - 1,\ u + v = 2y - 1,\ u - v = 2z - 1$$

and we obtain

$$8n + 3 = (2x - 1)^2 + (2y - 1)^2 + (2z - 1)^2$$

or

$$n = \frac{x^2 - x}{2} + \frac{y^2 - y}{2} + \frac{z^2 - z}{2}.$$

13. Yes, it should, in harmony with the pattern introduced in sect. 6.

No solution: **11, 12, 14.**

SOLUTIONS, CHAPTER XIII

1. The following pattern is generally applicable:

$$\begin{array}{c} A \text{ implies } B \\ B \text{ false} \\ \hline A \text{ false} \end{array}$$

Apply it in substituting non-B for B. You obtain

$$\begin{array}{c} A \text{ implies non-}B \\ \text{non-}B \text{ false} \\ \hline A \text{ false} \end{array}$$

We have noted in sect. 4 (5) the equivalence:

"*A* implies *B*" eq. "*A* incompatible with non-*B*".

Also this equivalence is generally applicable. Apply it in substituting non-*B* for *B*. You obtain

"*A* implies non-*B*" eq. "*A* incompatible with *B*"

We took here for granted that the negation of non-*B* is *B* (since the negation of *B* is non-*B*) and, therefore, we substituted *B* for non-(non-*B*). In substituting *B* for *A* in the equivalence noted at the end of sect. 4 (3), we obtain also

"non-*B* false" eq. "*B* true".

Substituting for the two premises of the last pattern the corresponding equivalent statements, displayed on the right-hand sides of the two foregoing equivalences, we obtain:

A incompatible with *B*
B true

A false

This is, in fact, the demonstrative pattern of sect. 3.

2. Assume that the following pattern is generally applicable:

A implies *B*
B true

A more credible

Apply it in substituting non-*A* for *A* and non-*B* for *B*. You obtain:

non-*A* implies non-*B*
non-*B* true
---
non-*A* more credible

Collect the following three equivalences:

"non-*A* implies non-*B*" eq. "*B* implies *A*"

"non-*B* true" eq. "*B* false"

"non-*A* more credible" eq. "*A* less credible".

The first has been derived in sect. 4 (5). The second has been mentioned in sect. 4 (3), in another notation, with *A* for *B*. The third has been stated (invented just for the present purpose) in sect. 5. Substitute for the premises and the conclusion of the last considered pattern the three corresponding equivalent statements just displayed. You obtain:

$$\frac{\begin{array}{c} B \text{ implies } A \\ B \text{ false} \end{array}}{A \text{ less credible}}$$

Except for a slight change in the wording (or notation) this is, in fact, the heuristic pattern introduced in sect. 2.

3. Start from the same pattern as in ex. 2. Substitute non-B for B (as in ex. 1). You obtain so:

$$\frac{\begin{array}{c} A \text{ implies non-}B \\ \text{non-}B \text{ true} \end{array}}{A \text{ more credible}}$$

Collect the following two equivalences:

"A implies non-B" eq. "A incompatible with B"

"non-B true" eq. "B false".

The first has been derived in ex. 1. The second is given, except for notation (A instead of B) in sect. 4 (3). Substitute for the two premises of the pattern considered the equivalent statements just displayed. You obtain

$$\frac{\begin{array}{c} A \text{ incompatible with } B \\ B \text{ false} \end{array}}{A \text{ more credible}}$$

This is, in fact, the heuristic pattern of sect. 3.

4. (a) At any rate A implies B, where the statement B is defined as follows: B. The letters in which the required nine-letter word is crossed by other words of the puzzle are chosen among the letters of the word TIREDNESS.

Let us interpret B as restricted to the two places filled in the proposed diagram (end and third from the end). We should distinguish between two cases.

If we regard the solutions for the two crossing words as final, we found that B is true and so we verified a consequence of the conjecture A. Therefore, we consider A as more credible according to the fundamental inductive pattern (sect. 12.1).

If, however, we regard the solutions for the crossing words only as tentative, we rendered B only more credible. Therefore the shaded version of the fundamental inductive pattern, defined in sect. 6, is appropriate and, of course, the evidence for A is weaker than in the former case.

(b) DISSENTER.

5. The pattern is that of sect. 13 (5):

> The Factum Probans is readily credible or understandable under assumption of the Factum Probandum.
>
> The Factum Probans is (much) less readily credible or understandable without the assumption of the Factum Probandum.
>
> The Factum Probans itself is proved.
>
> ---
>
> This renders the Factum Probandum more credible.

The presentation that we are considering here appears to be more appropriate to a court case, but the presentation of sect. 10 is more suitable to show the connection with the most usual form of inductive reasoning in the physical sciences or in mathematical research.

6. We have to treat the charge as a conjecture:

A. The down payment for the official's car came from the contractor's pocket.

We have to regard as a fact:

B. The withdrawal from the contractor's account of an amount ($875) equal to the down payment on the official's car, the date of the withdrawal preceding the date of the payment by two days.

B with *A* is much more readily understandable than *B* without *A*: if the withdrawal was not connected with the following payment, the exact coincidence of the amounts and the near coincidence of the dates has to be ascribed to mere chance. Such chance is not impossible, but improbable. The strength of the evidence hinges on this point. The pattern of sect. 13 (5) seems to fit very well.

7. Let us call poor Mrs. White and Mr. Black the "defendants." (They cannot answer charges, but at least they cannot be cross-examined by Mrs. Green.) Mrs. Green's accusation, stripped from her pious circumlocutions, is, of course,

A. The defendants live in double adultery.

We accept as a fact

B. The defendants had a long conversation in the obscurity over the fence.

This fact, unfortunately, yields some circumstantial evidence for *A* according to a reasonable pattern (sect. 13 (5)) if the following two premises are accepted:

B with *A* readily credible,

B without *A* less readily credible.

There is no use trying to shake the faith of Suburbia in the first premise, I am afraid. Yet even some Suburbans may see that, on that famous evening, the defendants might well have discussed the lease in which both had a legitimate interest. This makes B about as credible without A as with A, knocks out the second premise, and explains away the alleged circumstantial evidence. This argument may be of no avail against Mrs. Green's gossip, although it seems to me reasonable and typical. In arguing against some piece of circumstantial evidence, lawyers very often try to knock out just that second premise of the pattern in a "rebuttal."

8. We have to consider another contention or conjecture:

B. The defendant was well acquainted with the victim three years before the crime.

It would be too much to say that A is implied by B, but a weaker statement in this direction is obviously justified:

A is rendered more credible by B.

Now C does not prove B, but certainly renders B more credible.

From the two displayed premises we are tempted to draw the conclusion: A more credible.

This seems to suggest a new pattern:

$$\begin{array}{c} A \text{ more credible with } B \\ B \text{ more credible} \\ \hline A \text{ more credible} \end{array}$$

The first premise of this pattern is weaker than the corresponding premise of the pattern in line 2, column (2) of Table I, the second premise is the same: the conclusion must also be weaker. [See, however, ex. 15.2.]

By the way, the size of the firm matters: if the firm is small, B is rendered much more credible than in the opposite case.

No solution: Ex. **9** through **20**.

SOLUTIONS, CHAPTER XIV

1. (a) Follows from $r_r + s_r = r_s + s_s = 1$.
(b) $r_r - r_s = s_s - s_r > 0$.

2. With the notation of ex. 1 appropriately adapted, $r_r - r_s = s_s - s_r < 0$.

3. $N \binom{n}{s} p^s q^{n-s}$, with $N = 26306$, $n = 12$, $p = 1/3$, $q = 2/3$ where s is the number in the same row in column (1).

4. Same expression as in the solution of ex. 3 with the same numerical values for N and n and the same meaning of s, but $p = 0.3376986$, $q = 1 - p$.

5. (a) $N\mu^s e^{-\mu}/s!$ with $N = 30$, $\mu = 10$, and s the corresponding entry in column (1).
(b) $- - - + + - - + - - - + + + - + - + + +$.

6. 6^{-3n}.

7. The compound event that consists in casting six spots with each of the three dice five times in uninterrupted succession and has the probability 6^{-15} on the hypothesis of fair dice: $n = 5$ in ex. 6.

8. (a) $2 \int_{\alpha}^{\infty} y\, dx = 1.983 \cdot 10^{-7}$

where $y = (2\pi)^{-1/2} e^{-x^2/2}$, $\alpha = 1377.5\,(pqn)^{-1/2}$,

$$p = 1/3, \qquad q = 2/3, \qquad n = 315672.$$

(b) $\int_{\alpha}^{\beta} y\, dx = 1.506 \cdot 10^{-3}$

with $\beta = -\alpha = 0.5\,(pqn)^{-1/2}$ and y, p, q, and n have the same meaning as under (a). To compute the numerical values as exactly as they are given here the simplest current tables of the probability integral are not sufficient.

9. For finding no defective article in the sample, just one defective, precisely two defectives, . . . precisely c defectives, respectively.

10. $d^2a/dp^2 = 0$ if the derivative of log (da/dp) vanishes, which yields the equation

$$\frac{c}{p} - \frac{n - 1 - c}{1 - p} = 0$$

and hence the value given at the end of sect. 8 (1).

11. The required probability is 10^{-n}, *provided* that we accept one or the other of the following assumptions:

(I) All possible sequences of n figures are equally probable. (There are 10^n such sequences.)

(II) The various figures in the sequence are mutually independent, and for each figure the ten possible cases 0, 1, 2, . . . 9 are equally probable. (Apply the rule of sect. 3 (5) repeatedly.)

Both assumptions look "natural," but no such assumption is logically binding: the answer 10^{-n}, although *strongly suggested*, is *not* mathematically determined.

12. Suppose that l different letters can be drawn from a bag, with probabilities $p_1, p_2, \ldots p_l$, respectively. We have two such bags and we pick out a letter from each: the probability of a coincidence is $p_1^2 + p_2^2 + \ldots + p_l^2$. In the case of Hypothesis II, $l = 17$ and $p_1, \ldots p_{17}$ can be found by actual count.

13. In both cases

$$\sum_{k=n}^{10} \binom{10}{k} p^k q^{10-k} = 1 - \sum_{k=0}^{n-1} \binom{10}{k} p^k q^{10-k}$$

where $q = 1 - p$; (a) $p = 0.0948$, (b) $p = 1/26 = 0.03846$.

14. In both cases np with $p = 0.0948$; (a) $n = 450$, (*b*) $n = 90$.

15. $(npq)^{1/2}$ for $n = 90$, $p = 0.0948$, $q = 1 - p$. The computation of the standard deviation 7.60 is based on a formula which is not found in the textbooks. With the notation of ex. 12 set

$$p = p_1^2 + p_2^2 + \ldots + p_l^2,$$

$$p' = p_1^3 + p_2^3 + \ldots + p_l^3,$$

$$\sigma^2 = n(n-1)\, w[p(1-p) + 2(n-2)\,(p' - p^2)]/2.$$

Then $n = w = 10$, $p = 0.0948$, $p' = 0.01165$ yield $\sigma = 7.60$.

16. We assume that the 60 trials with the coin are independent and apply the rule introduced in sect. 3 (5) repeatedly.

17. Generalize the proposed numerical table

s	r	n
s'	r'	n'
S	R	N

and interpret it as follows. There are $N = R + S = n + n'$ cards, among which $R = r + r'$ cards are red and $S = s + s'$ cards are black. The cards are distributed at random between two players; one receives n cards, and the other n' cards. What is the probability that the first player receives r red cards and s black cards, and the other player r' red cards and s' black cards? (Of course, $r + s = n$, $r' + s' = n'$.) The answer is, as is well known,

$$\frac{\binom{R}{r}\binom{S}{s}}{\binom{N}{n}} = \binom{n}{r}\frac{R(R-1)\ldots(R-r+1)\,S(S-1)\ldots(S-s+1)}{N(N-1)\ldots(N-n+1)}$$

$$= \frac{1}{N!}\,\frac{R!\,S!\,n!\,n'!}{r!\,s!\,r'!\,s'!}.$$

Only four quantities out of the 9 contained in the table can be arbitrarily given, the values of the remaining 5 follow from the relations written above. We take the numbers $n = 9$, $n' = 11$, and $S = 8$ as given (the number of patients receiving each treatment and the total number of fatal cases) from which $N = 20$ and $R = 12$ follow. Yet we take for $s' = 2, 1, 0$ in succession (number of fatalities with the second treatment ≤ 2). From the formula, we compute the probability in each of these three cases (not more than 2 fatalities with the new treatment, 2, 1 or 0 black cards to Mr. Newman) and adding the probabilities for these mutually exclusive events we find:

$$\frac{12!\,8!\,9!\,11!}{20!}\left[\frac{1}{3!\,6!\,9!\,2!}+\frac{1}{2!\,7!\,10!\,1!}+\frac{1}{1!\,8!\,11!\,0!}\right]$$

$$=\left[\binom{12}{9}\binom{8}{2}+\binom{12}{10}\binom{8}{1}+\binom{12}{11}\binom{8}{0}\right]\Big/\binom{20}{11}=\frac{335}{8398}.$$

18. By an obvious extension of the reasoning of sect. 3 (5) (three-dimensional analogue of fig. 14.2) the number of possible cases is n^3. All the favorable cases, that is, all the admissible solutions of the equation $Z = X + Y$ can be enumerated as follows:

$$2 = 1 + 1$$

$$3 = 1 + 2 = 2 + 1$$

$$4 = 1 + 3 = 2 + 2 = 3 + 1$$

$$\cdot \quad \cdot \quad \cdot \quad \cdot$$

$$n = 1 + (n - 1) = 2 + (n - 2) = \ldots = (n - 1) + 1\ .$$

Hence the number of favorable cases is

$$1 + 2 + 3 + \ldots + (n - 1) = n(n - 1)/2$$

and the required probability

$$\frac{n(n-1)/2}{n^3} = \frac{n-1}{2n^2}.$$

21. The probability that a sample of 38 from an infinite population contains 30 or more defectives is

$$\sum_{i=30}^{38}\binom{38}{i}p^i(1-p)^{38-i} < \binom{38}{8}p^{30}(1-p)^8\sum_{i=0}^{\infty}\left(\frac{p}{1-p}\right)^i$$

$$= \binom{38}{8}\frac{(1-p)^9}{1-2p}p^{30}$$

$$\sim 4.61 \times 10^{-23}$$

provided that $100p\% = 1\%$ of the population is defective. The probability estimated is the likelihood of the daily's assertion in the light of the official's observation, computed under the simplest assumptions.

22. The probability required is

$$4\left\{\left(\frac{2}{12\times 60}\right)^3+3\left(\frac{2}{12\times 60}\right)^2\left(1-\frac{2}{12\times 60}\right)\right\}\sim 0.0000924.$$

(Any one of the four clocks could be the one among the three agreeing clocks that shows the earliest time. The possibility that all four clocks show times less than 2 minutes apart accounts for the first term in the curly brackets.)

23. The integer a takes one of the values

$$-n, \ldots, -2, -1, 0, 1, 2 \ldots n.$$

We assume that these $2n + 1$ values are equally probable, we make the corresponding assumption for b, c, d, e, and f, and assume also that a, b, c, d, e, and f are mutually independent. Only now, after having given a precise meaning to "chosen at random," can we proceed to solve the problem.

There is just one solution if, and only if, $ad - bc \neq 0$. We can, and do, neglect e and f: there are $(2n + 1)^4$ possible cases. We count *unfavorable* cases, distinguishing two possibilities.

(I) $a = 0$. Then $bc = 0$, d arbitrary, and there are $(2n + 2n + 1)(2n + 1)$ cases.

(II) $a \neq 0$. Then a can take $2n$ values and d is uniquely determined by a, b, and c: there are at most $2n(2n + 1)^2$ cases.

The probability required is

$$\geqq 1 - \frac{(4n + 1)(2n + 1) + 2n(2n + 1)^2}{(2n + 1)^4}$$

$$= 1 - \frac{4n^2 + 6n + 1}{8n^3 + 12n^2 + (6n + 1)} > 1 - \frac{1}{2n}$$

and so it tends to 1 as n tends to ∞. This gives another precise meaning to the statement: "a system of two equations with two unknowns has *in general* just one solution." Cf. sect. 11.3, ex. 11.16.

24. The verification of the O gives more confidence in TOWER than the verification of the E. Since O is less frequent than E, the occurrence of O in the crossing word is less easily interpreted as chance coincidence than that of the E.

25. Tabulating the differences between the successive numbers of each column, and then again the differences of the differences (the so-called "second differences") we obtain:

I			II		
1005			1004		
	28			34	
1033		+ 14	1038		0
	42			34	
1075		− 11	1072		0
	31			34	
1106		− 5	1106		− 1
	26			33	
1132		+ 21	1139		+ 1
	47			34	
1179		− 21	1173		− 1
	26			33	
1205		0	1206		0
	26			33	
1231		+ 17	1239		− 1
	43			32	
1274		− 16	1271		0
	27			32	
1301			1303		

The first differences show it clearly enough that II is regular and I is not. Yet the second differences are still more suggestive: II shows a minimum of irregularity due to the unavoidable rounding errors, but the second differences in I vary in sign and are quite large. Such "differencing" is an important operation in checking the construction of numerical tables. There are two remarks.

(1) In the table of a function $f(x)$ the first differences are connected with $f'(x)$, and the second differences with $f''(x)$, by the mean value theorem. This provides us with an opportunity to check the differences.

(2) The last decimals in I are, in fact, the first ten decimals of π in reverse order. The view that the successive decimals of π behave *as if* they were produced by chance, has been expressed many times with many variations.

27. Assume that A is independent of B:

$$\Pr\{A/B\} = \Pr\{A/\bar{B}\}. \tag{6}$$

(We begin with (6) to avoid interference with the numbering in ex. 26.)

Using (4), (2), (6), and (3) (in this order) we obtain

$$\begin{aligned}\Pr\{A\} &= \Pr\{AB\} + \Pr\{A\bar{B}\} \\ &= \Pr\{B\}\Pr\{A/B\} + \Pr\{\bar{B}\}\Pr\{A/\bar{B}\} \\ &= \Pr\{A/B\}\,(\Pr\{B\} + \Pr\{\bar{B}\}) \\ &= \Pr\{A/B\}.\end{aligned} \tag{7}$$

From (2), (7), and $\Pr\{A\} \neq 0$ follows that

$$\begin{aligned}\Pr\{A\}\Pr\{B/A\} &= \Pr\{B\}\Pr\{A\} \\ \Pr\{B/A\} &= \Pr\{B\}.\end{aligned} \tag{8}$$

Using again (4) and (2), as in (7), and using also (8), (3), and $\Pr\{\bar{A}\} \neq 0$, we find that

$$\begin{aligned}\Pr\{B\} &= \Pr\{A\}\Pr\{B/A\} + \Pr\{\bar{A}\}\Pr\{B/\bar{A}\} \\ (1 - \Pr\{A\})\Pr\{B\} &= \Pr\{\bar{A}\}\Pr\{B/\bar{A}\} \\ \Pr\{B\} &= \Pr\{B/\bar{A}\};\end{aligned} \tag{9}$$

(6), (7), (8), and (9) show the required conclusion.

28. If A and B are mutually independent,

$$\Pr\{AB\} = \Pr\{A\}\Pr\{B\}.$$

This follows from rule (2) of ex. 26 and definition (II) of ex. 27.

29. (a)

	$\Pr\{A\}$,	$\Pr\{A/B\}$,	$\Pr\{A/\bar{B}\}$,	$\Pr\{B\}$,	$\Pr\{B/A\}$,	$\Pr\{B/\bar{A}\}$
(I)	$\frac{1}{3}$,	$\frac{1}{3}$,	$\frac{1}{3}$,	$\frac{1}{2}$,	$\frac{1}{2}$,	$\frac{1}{2}$
(II)	$\frac{1}{3}$,	1,	$\frac{1}{5}$,	$\frac{1}{6}$,	$\frac{1}{2}$,	0

(b)

$$\Pr\{AB\} = \Pr\{A\}\Pr\{B/A\} = \Pr\{B\}\Pr\{A/B\}$$

(I) $\quad \frac{1}{3} \cdot \frac{1}{2} = \frac{1}{2} \cdot \frac{1}{3}$

(II) $\quad \frac{1}{3} \cdot \frac{1}{2} = \frac{1}{6} \cdot 1$.

(c) The formulas follow generally from ex. 26 (4), (2). Numerically

$$\Pr\{A\} = \Pr\{B\}\Pr\{A/B\} + \Pr\{\bar{B}\}\Pr\{A/\bar{B}\},$$

(I) $\quad \frac{1}{3} = \frac{1}{2} \cdot \frac{1}{3} + \frac{1}{2} \cdot \frac{1}{3}$,

(II) $\quad \frac{1}{3} = \frac{1}{6} \cdot 1 + \frac{5}{6} \cdot \frac{1}{5}$.

$$\Pr\{B\} = \Pr\{A\}\Pr\{B/A\} + \Pr\{\bar{A}\}\Pr\{B/\bar{A}\},$$

$$\text{(I)} \quad \tfrac{1}{2} = \tfrac{1}{3} \cdot \tfrac{1}{2} + \tfrac{2}{3} \cdot \tfrac{1}{2},$$

$$\text{(II)} \quad \tfrac{1}{6} = \tfrac{1}{3} \cdot \tfrac{1}{2} + \tfrac{2}{3} \cdot 0 .$$

(d) With (I), A and B are mutually independent, with (II) they are not.

No solution: **19**, **20**, **26**, **30**, **31**, **32**, **33**.

SOLUTIONS, CHAPTER XV

1. Since H implies both A and B,

$$\Pr\{A/H\} = 1, \qquad \Pr\{B/H\} = 1.$$

Therefore, by rule (2) of ex. 14.26,

$$\Pr\{H\} = \Pr\{A\}\Pr\{H/A\}, \qquad \Pr\{H\} = \Pr\{B\}\Pr\{H/B\}.$$

Eliminating $\Pr\{H\}$, we obtain

$$\Pr\{A\}\Pr\{H/A\} = \Pr\{B\}\Pr\{H/B\}.$$

If we regard the relation between H and A, and also the relation between H and B, as unchanged, and, therefore, $\Pr\{H/A\}$ and $\Pr\{H/B\}$ as constants, and let $\Pr\{B\}$ increase, also $\Pr\{A\}$ will increase by virtue of our last equation.

2. By the formulas ex. 14.26 (2), (3), (4)

$$\Pr\{A\} = \Pr\{A/\bar{B}\} + \Pr\{B\}\left(\Pr\{A/B\} - \Pr\{A/\bar{B}\}\right).$$

We take $\Pr\{A/B\}$ and $\Pr\{A/\bar{B}\}$ as given, but let $\Pr\{B\}$ increase. (This can be quite naturally visualized in the case considered in ex. 13.8.) The increase of $\Pr\{B\}$ implies a corresponding increase of $\Pr\{A\}$ *if*

$$(*) \qquad \Pr\{A/B\} > \Pr\{A/\bar{B}\}.$$

In the concrete case of ex. 13.8 this inequality seems acceptable. Yet the statement of the pattern given in the solution of ex. 13.8 appears unacceptable. Here is a statement which looks better and is certainly in line with the formula:

$$\frac{\begin{array}{c} A \text{ more credible with } B \text{ than without } B \\ B \text{ (becomes) more credible} \end{array}}{A \text{ (becomes) more credible}}$$

If A is a consequence of B, $\Pr\{A/B\} = 1$, (*) is certainly correct, and the pattern becomes that in sect. 13.7, Table I, line 2, column 2.

3. Review the few cases in which we have represented patterns of plausible reasoning by formulas of the Calculus of Probability: sect. 6–10 and ex. 1–2.

The monotonicity and continuity asserted by ex. 13.10 is obvious in these cases from the following simple mathematical fact: If a, b, c, d, x_1, x_2 are real constants, $ad - bc \neq 0$, and the function y of x, defined by

$$y = \frac{ax + b}{cx + d},$$

has the property that $0 \leqq y \leqq 1$ for $x_1 < x < x_2$, then y is strictly monotonic and continuous for $x_1 \leqq x \leqq x_2$. With respect to the generality of the remarks of ex. 13.10 we should observe this: if y is not a linear fractional function but, more generally, a rational function of x, it is still necessarily a continuous function of x, except at points where it becomes infinite, but no more necessarily a monotonic function.

No solution: **4**, **5**, **6**, **7**, **8**, **9**.

SOLUTIONS, CHAPTER XVI

2. [Stanford 1951] The quadrilateral has to be convex. Let us call I, II, III, IV the triangles into which it is divided by its diagonals, (I), (II), (III), (IV) the areas of the four triangles, respectively, and p, q, r, s the lengths of the four straight lines drawn from the intersection of the diagonals to the four vertices of the quadrilateral. Name and number in "cyclic order" so that the side of length p is common to IV and I, q to I and II, r to II and III, s to III and IV; I is opposite to III, II to IV; $p + r$ is the length of one diagonal, $q + s$ that of the other. Let p and q include the angle α. Then

$$2(\text{I}) = pq \sin \alpha, \qquad 2(\text{II}) = qr \sin \alpha,$$

$$2(\text{III}) = rs \sin \alpha, \qquad 2(\text{IV}) = sp \sin \alpha.$$

Hence

(a) (I) (III) = (II) (IV).

(b) The base of I is parallel to that of III if, and only if,

$$p/q = r/s \qquad \text{or} \qquad (\text{II}) = (\text{IV}).$$

(c) The quadrilateral is a parallelogram if, and only if,

$$p = r, \quad q = s \quad \text{or} \quad (\text{I}) = (\text{II}) = (\text{III}) = (\text{IV}).$$

3. [Stanford 1949] (a) Let a be a side of the equilateral triangle. Joining the point inside it to its three vertices, you divide it into three triangles with areas that added together give the whole area: $ax/2 + ay/2 + az/2 = ah/2$. Divide by $a/2$. See fig. 8.8.

(b) A point inside a regular tetrahedron with altitude h has the distances x, y, z, and w from the four faces, respectively. Then $x + y + z + w = h$. The proof is analogous: divide the regular tetrahedron into four tetrahedra.

(c) The relation remains valid in both cases (a) and (b) for outside points, provided that the distances x, y, z (and w) are taken with the proper sign: $+$ when a spectator placed in the point sees the side (face) from inside, $-$ when he sees it from outside. The proof is essentially the same.

4. [Stanford 1946] (a) (I) and (IV) are generally true, but (II) and (III) are not; see (b).

(b) (II) and (III) are false: the rectangle and the rhombus are counterexamples, respectively.

(c) (II) and (III) are true for pentagons.

(d) (II) and (III) are true for polygons with an odd number of sides, 3 or 5 or 7 . . . , as it follows from (II′) and (III′).

(II′) If a polygon inscribed in a circle is equiangular, any two sides separated by just one intervening side are equal. Therefore, if the number of sides is even, equal to $2m$, either all $2m$ sides are equal, or m sides are equal to a and the m remaining sides to b, $a \neq b$, and no two sides having a vertex in common are equal.

(III′) If a polygon circumscribed about a circle is equilateral, any two angles separated by just one intervening angle are equal. Therefore, if the number of angles is even, equal to $2m$, either all $2m$ angles are equal, or m angles are equal to α and the m remaining angles to β, $\alpha \neq \beta$, and no two angles having a side in common are equal.

To prove (I) (II′) (III′) (IV) join the center of the circle to the vertices of the polygon, draw perpendiculars from the center to the sides, and pick out congruent triangles.

5. [Stanford 1947] The relation between (a) and (b) appears similar to that between (b) and (c); the similarity of the left-hand sides appears more clearly than that of the right-hand sides. Hence it is natural to look for some transition from (a) to (b) that could also serve as a transition from (b) to (c). There is such a transition: if

$$\alpha + \beta + \gamma = \pi$$

$$2\alpha + 2\beta + 2\gamma = 2\pi \neq \pi, \text{ but}$$

$$(\pi - 2\alpha) + (\pi - 2\beta) + (\pi - 2\gamma) = \pi.$$

Taking for granted that (a) holds for any three angles α, β, γ with sum π, we obtain (b) by substituting for these angles $\pi - 2\alpha$, $\pi - 2\beta$, $\pi - 2\gamma$, respectively, and we pass from (b) to (c) by using the same substitution.

It remains to verify (a) which can be done in many ways, for instance as follows. Substituting $2u$, $2v$, and $\pi - 2u - 2v$ for α, β, and γ, respectively, we transform (a) into

$$\sin u \cos u + \sin v \cos v = [2 \cos u \cos v - \cos (u + v)] \sin (u + v).$$

Use the addition theorems of cosine and sine.

6. [Stanford 1952] According to the four proposals, the volume of the frustum would be

$$\text{(I)}\ [(a+b)/2]^2h$$

$$\text{(II)}\ [(a^2+b^2)/2]h$$

$$\text{(III)}\ [a^2+b^2+(a+b)^2/4]h/3$$

$$\text{(IV)}\ [a^2+b^2+ab]h/3,$$

respectively. If $b = a$, the frustum becomes a prism with volume a^2h: (I) (II) (III) (IV) agree in yielding the correct result. If $b = 0$ the frustum becomes a pyramid with volume $a^2h/3$: only (IV) yields this and so (I) (II) (III) which yield different values must be incorrect. That (IV) is generally correct, must still be proved; see the textbooks.

10. (1) Doing the addition with paper and pencil in the most usual manner, you begin at the top of the last column and proceed downward; in the given example the first two steps are $3 + 7 = 10$, $10 + 0 = 10$. Then you proceed downward in the next column, and so on. (2) You take the columns in the same order as before, but you start at the bottom and proceed upward in each. (3) You take the columns again in the same order but do each column twice, first downward, then upward; you note the result when, working downward, you reach the bottom, and check it by working upward. (4) Add first the 1st, 3rd, 5th, . . . number, then the 2nd, 4th, 6th, . . . number, and finally the two sums obtained; this requires some extra writing. (5) Do the addition once by hand, and once by a machine. And so on.

11. 59.

12. 10^{-9}. (Not 10^{-3} or 10^{-7}.)

No solution: **1**, **7**, **8**, **9**, **13**.

BIBLIOGRAPHY

I. CLASSICS

EUCLID, *Elements.* The inexpensive shortened edition in Everyman's Library is sufficient here. "EUCLID III, 7" refers to Proposition 7 of Book III of the Elements.

DESCARTES, *Oeuvres,* edited by Charles Adam and Paul Tannery. The work "Regulae ad Directionem Ingenii," vol. 10, pp. 359–469, is of especial interest.

EULER, *Opera Omnia,* edited by the "Societas scientiarum naturalium Helvetica."

LAPLACE, *Oeuvres complètes.* The "Introduction" of vol. 7, pp. V–CLIII, also separately printed (and better known) under the title "Essai philosophique sur les probabilites" is of especial interest.

II. SOME BOOKS OF SIMILAR TENDENCY

R. COURANT and H. ROBBINS, *What is mathematics?*

H. RADEMACHER and O. TOEPLITZ, *The enjoyment of mathematics*, Princeton, 1957.

O. TOEPLITZ, *The calculus,* Chicago, 1963; of especial interest.

III. RELATED FORMER WORK OF THE AUTHOR

Books

1. *Aufgaben und Lehrsätze aus der Analysis*, 2 volumes, Berlin, 1925. Jointly with G. Szegö. Third corrected edition 1964.
2. *How to Solve It*, Princeton, 1945. Second (paperback) edition, Garden City, 1957.
3. Wahrscheinlichkeitsrechnung, Fehlerausgleichung, Statistik. From *Abderhalden's Handbuch der biologischen Arbeitsmethoden*, Abt. V, Teil 2, pp. 669–758.

Papers

1. Geometrische Darstellung einer Gedankenkette. *Schweizerische Pädagogische Zeitschrift*, 1919, 11 pp.
2. Wie sucht man die Lösung mathematischer Aufgaben? *Zeitschrift für mathematischen und naturwissenschaftlichen Unterricht*, v. 63, 1932, pp. 159–169.
3. Wie sucht man die Lösung mathematischer Aufgaben? *Acta Psychologica*, v. 4, 1938, pp. 113–170.
4. Heuristic reasoning and the theory of probability. *American Mathematical Monthly*, v. 48, 1941, pp. 450–465.

5. On Patterns of Plausible Inference. *Studies and Essays presented to R. Courant*, 1948, pp. 277–288.

6. Generalization, Specialization, Analogy. *American Mathematical Monthly*, v. 55, 1948, pp. 241–243.

7. Preliminary remarks on a logic of plausible inference. *Dialectica*, v. 3, 1949, pp. 28–35.

8. With, or without, motivation? *American Mathematical Monthly*, v. 56, 1949, pp. 684–691.

9. Let us teach guessing. *Etudes de Philosophie des Sciences, en hommage à Ferdinand Gonseth*, 1950, pp. 147–154. Editions du Griffon, Neuchatel, Switzerland.

10. On plausible reasoning. *Proceedings of the International Congress of Mathematicians*, 1950, v. 1, pp. 739–747.

IV. PROBLEMS

Among the examples proposed for solution there are some taken from the *William Lowell Putnam Mathematical Competition* or the *Stanford University Competitive Examination in Mathematics*. This fact is indicated at the beginning of the solution with the year in which the problem was proposed as "Putnam 1948" or "Stanford 1946." The problems of the Putnam Examination are published yearly in the *American Mathematical Monthly* and most Stanford examinations have been published there too.

[The related work of the author, *Mathematical Discovery* (New York, London, Sydney; vol. I, 1962; vol. II, 1965) contains additional bibliographical data; see vol. II, pp. 184–186.]

V. FILM

Let us teach guessing. A demonstration with George Polya, presented by a committee of the Mathematical Association of America.

Appendix

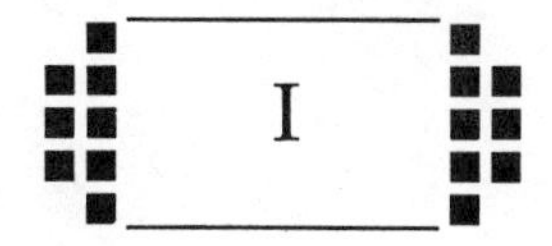

HEURISTIC REASONING IN THE THEORY OF NUMBERS[1]

A deep but easily understandable problem about prime numbers is used in the following to illustrate the parallelism between the heuristic reasoning of the mathematician and the inductive reasoning of the physicist. The experts may judge whether the parallelism is more serious than the tone of presentation which is adapted to a wider audience.

1. "Till now the mathematicians tried in vain to discover some order in the sequence of the prime numbers and we have every reason to believe that there is some mystery which the human mind shall never penetrate. To convince oneself, one has only to glance at the tables of primes which some people took the trouble to compute beyond a hundred thousand, and one perceives that there is no order and no rule. This is so much more surprising as the arithmetic gives us definite rules with the help of which we can continue the sequence of the primes as far as we please, without noticing, however, the least trace of order."[2]

So wrote Euler about two centuries ago, yet the prime numbers may inspire the contemporary mathematician with the same feeling of mystery that Euler so vividly expressed. The primes remain puzzling in spite of many important discoveries made in the meantime. Let us look at some of these discoveries.

The intervals between successive primes are irregular, but these intervals seem to become larger "on the whole" (the primes seem to become scarcer) as we proceed in the sequence of numbers. Since Euler's time a definite law of this phenomenon was discovered (conjectured by Legendre and Gauss, investigated by Chebyshev and Riemann, finally proved by Hadamard and de la Vallée Poussin, proved recently in an essentially different "elementary" manner by Atle Selberg and Paul Erdös). We may formulate this law, the "prime number theorem," intuitively although not quite precisely, as

---

[1] Reprinted from the *American Mathematical Monthly* vol. 66, 1959, pp. 375–384, with the permission of The Mathematical Association of America.

[2] See L. Euler, *Opera Omnia*, ser. 1, vol. 2, p. 241 or sect. 6.2 of this work.

follows: The probability that a large integer x should be a prime, is $1/\log x$ (where $\log x$ is the natural logarithm of x).[3]

The following short table exhibits the first primes (with two exceptions) classified according to their last digit.

	11		31	41		61	71			101	
3	13	23		43	53		73	83		103	113
7	17		37	47		67			97	107	
	19	29			59		79	89		109	

If we set apart 2 and 5, the prime factors of 10, the last figure in the decimal symbol of a prime cannot be 0, 2, 4, 5, 6, or 8 (since neither 2 nor 5 should be a divisor) and must, therefore, be 1, 3, 7, or 9. Thus, with respect to ten (*modulo* 10) there are four kinds of primes which are listed in the four horizontal lines of the foregoing table, respectively. Since Euler's time, a general law has been discovered (most of the credit for its discovery is due to Dirichlet) which, applied to our particular case, asserts that there are infinitely many prime numbers of each kind and, what is more, that each kind is equally probable. Therefore, in an extensive table of prime numbers there must be roughly as many primes ending with 1 as primes ending with 3.

Euler mentions a table of primes that goes beyond 10^5. Since his time much more extensive tables have been computed, especially in the last decade with the help of machines. Data derived from these tables may suggest problems not yet considered by Euler.

2. The least possible distance between two consecutive primes is 2, if we set apart the unique case of the primes 2 and 3. Two primes having this minimum distance are called *twin primes*. Here is a list of the twin primes under 100:

3, 5   5, 7   11, 13   17, 19   29, 31   41, 43   59, 61   71, 73

We can generalize this situation and consider a prime p that is escorted at a given distance d by another prime $p' = p + d$. (This situation is uninter-

[3] The irregular distribution of primes ("there is no order and no rule") strongly suggests the idea of probability and chance. Yet this is paradoxical: Whether any given integer is a prime or not, can be decided by the "definite rules" of arithmetic—where and how could chance enter the picture? The paradox can be somewhat explained (or deepened) by a physical analogy. The kinetic theory of matter considers the probability distribution of the velocities of the molecules in a gas. Yet this is paradoxical: The velocities resulting from the collision of two molecules can be exactly predicted from the data of the collision by the "definite rules" of classical deterministic mechanics—where and how could chance enter the picture? The determinateness of the simple single event and the probabilistic theory of the highly composite whole may seem to be equally compatible (or incompatible) in both cases.

esting unless d is even; we do not care whether there are or are not other primes between p and p'.) Here is a list of all such pairs at the distance 6, in which the first prime does not (but its escort may) exceed 100:

5, 11	7, 13	11, 17	13, 19	17, 23	23, 29	31, 37	37, 43
41, 47	47, 53	53, 59	61, 67	67, 73	73, 79	83, 89	97, 103

It is curious that the second kind of pairs is more numerous. We count 8 pairs of twin primes and exactly twice as many pairs of primes at the distance 6. Let us take now instead of 10^2 the considerably higher bound $3 \cdot 10^7$. Under thirty million there are 152892 primes followed by another prime at the distance 2, but nearly twice as many, namely 304867 primes followed by another at the distance 6.

The numbers of these prime pairs have been obtained by Professor and Mrs. D. H. Lehmer with the use of appropriate computing apparatus; they computed, up to the same limit $3 \cdot 10^7$, the number of primes escorted by another prime at the distance d for $d = 2, 4, 6, 8, \ldots, 70$. I wish to thank them here for their kind permission to use their interesting material. I wish to use some of their results to offer the unprejudiced reader a particularly suitable opportunity for an inductive investigation in pure mathematics.

It will be convenient to introduce here some notation. Let $\pi_d(x)$ stand for the number of those prime numbers p that satisfy two conditions:

$$p \leqq x, \qquad p + d \text{ is a prime number.}$$

For instance,

$$\pi_2(100) = 8, \qquad \pi_2(30\ 000\ 000) = 152892,$$

$$\pi_6(100) = 16, \qquad \pi_6(30\ 000\ 000) = 304867.$$

I set

$$\pi_d(3 \cdot 10^7)/\pi_2(3 \cdot 10^7) = R_d.$$

For instance, $R_6 = 304867/152892 = 1.9940$, approximately. A small part of the material computed by Professor and Mrs. Lehmer is collected in Table I.

Table I. Values of R_d

d	R_d	d	R_d	d	R_d	d	R_d	d	R_d	d	R_d
		12	1.9985	24	1.9976	36	1.9997	48	1.9965	60	2.6632
2	1.0000	14	1.1985	26	1.0910	38	1.0566	50	1.3308	62	1.0341
4	0.9979	16	1.0001	28	1.1974	40	1.3330	52	1.0892	64	0.9999
6	1.9940	18	1.9982	30	2.6632	42	2.3987	54	1.9981	66	2.2186
8	0.9996	20	1.3311	32	0.9970	44	1.1097	56	1.1957	68	1.0663
10	1.3317	22	1.1088	34	1.0645	46	1.0467	58	1.0349	70	1.5977

3. Now, let us start our inductive research. At any moment at which the reader feels inspired, he should interrupt the reading and try to guess the result by himself.

The four kinds of prime numbers that we have considered in Section 1 (ending with 1, 3, 7 or 9 in the decimal notation, respectively) are known to be equally frequent. Are the 35 kinds of prime numbers with which Table I is concerned also equally frequent? If it were so, all the ratios R_d contained in Table I should be approximately equal to one. In fact, remarkably enough, a few entries in Table I are pretty close to the value 1, but the majority seem to deviate significantly from 1. The analogy with the previous case does not seem to go far. Yet, perhaps, the analogy holds at least in one respect: the ratio $\pi_d(x)/\pi_2(x)$ may converge towards some limit (not necessarily 1) when x tends to infinity, and the ratio $R_d = \pi_d(3 \cdot 10^7)/\pi_2(3 \cdot 10^7)$ entered into Table I may be an approximation to that limit.

We face here a situation somewhat analogous to the situation that the chemists faced around 1800 when they were about to discover the Law of Multiple Proportions. They had to perceive behind their experimental data distorted by unavoidable errors of observation the ratios of simple multiples of the atomic weights, and we have to perceive behind the approximate ratios R_d collected in Table I the true limiting ratios. To guess these limiting ratios is a challenging task.

We have already observed that some values of R_d are very close to 1; they correspond to $d = 2, 4, 8, 16, 32, 64$. (For $d = 2$ the value is exactly 1, but this is trivial.) We can scarcely fail to notice here the powers of 2. By the way, these values of R_d so close to 1 are also the smallest values in the table. Are there other entries in the table so nearly equal to each other?

In trying to answer this question we may notice that the entries corresponding to

$$d = 6, 12, 24, 48$$

are approximately equal to each other, and so are those corresponding to

$$d = 10, 20, 40$$

or those corresponding to

$$d = 14, 28, 56.$$

In general, multiplication of d by 2 seems to leave the value of R_d almost unchanged.

What about multiplication by 3? It approximately doubles the value of R_d in certain transitions, as from

2 to 6,	4 to 12,	8 to 24,	16 to 48,
10 to 30,	20 to 60,	14 to 42,	22 to 66.

Yet it is not so in other cases, as

$$6 \text{ to } 18, \qquad 12 \text{ to } 36, \qquad 18 \text{ to } 54;$$

in these latter cases the multiplication of d by 3 leaves the value of R_d almost unchanged. How can you account for this different behavior?

And so on, from question to question, by observation and tentative generalization, carefully checking each guess, the reader may discover that many of the values R_d contained in Table I come very close to simple fractions; see Table II.

Table II. Simple approximations to some R_d

d	2, 4, 8, 16, 32, 64	6, 12, 18, 24, 36, 48, 54	10, 20, 40, 50	14, 28, 56	22, 44	30, 60	42	66	70
R_d (approx.)	1	$\frac{2}{1}$	$\frac{4}{3}$	$\frac{6}{5}$	$\frac{10}{9}$	$\frac{8}{3}$	$\frac{12}{5}$	$\frac{20}{9}$	$\frac{8}{5}$

Table II strongly suggests that R_d *depends only on the decomposition of d into prime factors.* More precisely, just the presence of a prime factor in, or its absence from, the decomposition seems to be relevant; for instance, to all values of d of the form $2^\alpha 3^\beta$ with $\alpha, \beta = 1, 2, 3, \ldots$ there corresponds the same value of R_d (approximately).

Moreover, to each prime factor of d there seems to correspond a factor of R_d; to the (unavoidable) factor 2 of d, the (trivial) factor 1 of R_d; to the prime factors

$$3, \quad 5, \quad 7, \quad 11$$

of d, the following factors of R_d:

$$\frac{2}{1}, \quad \frac{4}{3}, \quad \frac{6}{5}, \quad \frac{10}{9},$$

respectively. Then, when d is a product of different primes (or powers of different primes) R_d seems to be the product of the corresponding factors.

4. All such observations point to the (conjectural) formula

$$\pi_d(x) \sim \pi_2(x) \prod_{p|d} \frac{p-1}{p-2}, \tag{1}$$

where the product $\prod_{p|d}$ is extended over all different *odd* prime factors p of the even number d.[4] The sign $\sim$ can be interpreted either vaguely or strictly. In a vague interpretation $\sim$ means "approximately equal;" in the strict sense it means "the ratio of the two sides tends to 1 when x tends to ∞." The formula is merely a conjecture which we can conceive quite naively by examining Table I. In Table III, the observed values of R_d, taken from Table I and styled now R_d (obs.), are compared with the corresponding conjectural limiting values, styled R_d (theor.). This comparison yields strong inductive evidence for the conjecture which could be further strengthened by use of other data computed by Professor and Mrs. Lehmer.

Table III. Values of R_d, observed and "theoretical"

d	R_d (obs.)	R_d (theor.)	24	1.9976	2.0000	48	1.9965	2.0000
2	1.0000	1.0000	26	1.0910	1.0909	50	1.3308	1.3333
4	0.9979	1.0000	28	1.1974	1.2000	52	1.0892	1.0909
6	1.9940	2.0000	30	2.6632	2.6667	54	1.9981	2.0000
8	0.9996	1.0000	32	0.9970	1.0000	56	1.1957	1.2000
10	1.3317	1.3333	34	1.0645	1.0667	58	1.0349	1.0370
12	1.9985	2.0000	36	1.9997	2.0000	60	2.6632	2.6667
14	1.1985	1.2000	38	1.0566	1.0588	62	1.0341	1.0345
16	1.0001	1.0000	40	1.3330	1.3333	64	0.9999	1.0000
18	1.9982	2.0000	42	2.3987	2.4000	66	2.2186	2.2222
20	1.3311	1.3333	44	1.1097	1.1111	68	1.0663	1.0667
22	1.1088	1.1111	46	1.0467	1.0476	70	1.5977	1.6000

5. We have before us a precise, general, but enigmatic formula derived from, and quite well verified by, observations. Of course, we wish to understand it, we wish to explain it. When we are looking at it, our situation is similar to that of Newton looking at the laws of Kepler or to that of Niels Bohr looking at Balmer's formula. The word "similar" must be correctly understood. Similar figures may be very different in magnitude, but they show the same proportions, and so do in a sense the three situations we have ust compared.

We wish to explain that conjectural formula about prime numbers. Both

[4] The usual abbreviation $a|b$ means "a divides b" or "a is a divisor of b." We shall need later also the abbreviation $a \nmid b$ which means "a is not a divisor of b."

the irregular distribution of the primes and the structure of the conjectural formula strongly suggest an explanation by probability. I wish to present such an explanation. We shall arrive at it in two steps (of which the second is much more dangerous).

PROBLEM I. *Let p denote a given prime number, d a given integer and x a large integer chosen at random. Find the probability that neither x nor $x + d$ is divisible by p.*

The reader may visualize the integers as successive intervals of equal length along an infinite straight line, some sort of super-roulette. The interval is red or green, according as the integer is, or is not, divisible by p; among any p consecutive intervals there is always just one that is red. A ball is rolled along the line and stops in the interval x.

We have to distinguish two cases.[5]

First case: $p \mid d$. In this case $x + d$ falls on a multiple of p (a red space) if, and only if, x itself falls on such a multiple. Therefore, out of any p consecutive numbers (spaces), $p - 1$ are favorable (green) and so the required probability is $(p - 1)/p$.

Second case: $p \nmid d$. Even if x does not fall on a multiple of p, $x + d$ may. Therefore, out of any p consecutive numbers just $p - 2$ are favorable. The required probability is $(p - 2)/p$.

PROBLEM II. *Let d denote a given even integer, and x a large integer chosen at random. Find the probability P_d that both x and $x + d$ are prime numbers.*

In order that both x and $x + d$ should be prime numbers, a sequence of conditions must be satisfied:

First, neither x nor $x + d$ is divisible by 2;
then, neither x nor $x + d$ is divisible by 3;
then, neither x nor $x + d$ is divisible by 5;
and so on. The general form of this condition is: neither x nor $x + d$ is divisible by p where p is a prime number.

We have computed above the probability for the fulfillment of any single one of these conditions. Now we have to compute the probability that all these conditions are fulfilled at the same time, all these events are realized simultaneously.

Two difficulties arise here: Are these events independent? How far should we go with p? In fact, these two difficulties may be connected, but at this stage of the game it will be better not to examine them too thoroughly; let us now proceed quickly and see whether anything worthwhile turns up.

Are the events independent? We do not know, but let us assume it. Also the physicist is inclined to assume the independence of the probabilities he deals with—not because he knows that they are independent, but inter-

---

[5] For the symbols $\mid$ and $\nmid$, see footnote 4.

dependent probabilities are so much more difficult to handle—and so let us assume independence in our case too, although we have no better reasons than the physicist.

Having made this assumption all we have to do is to multiply probabilities computed above. We distinguish three cases:

$p = 2$ (which is a divisor of the even number d);

p is odd and is a divisor of d;

p is odd and is not a divisor of d.

Accordingly, the required probability P_d is a product of three kinds of factors:

$$P_d = \frac{1}{2} \prod_{p|d} \frac{p-1}{p} \prod_{p \nmid d} \frac{p-2}{p}. \tag{2}$$

In this formula (2) (and in the following formulas (3), (4)) the letter p stands for an *odd* prime number.

How far should we go with p? Of course, on the right hand side of formula (2) we extend the first product over all odd prime factors of the given number d. In the second product, we take all the odd primes not dividing d up to a certain large upper bound, depending on the considered large number x—but let us *postpone* the decision, how far to go, how large that upper bound should precisely be.

We can transform formula (2) as follows:

$$P_d = \prod_{p|d} \frac{p-1}{p-2} \cdot \frac{1}{2} \prod_{p} \frac{p-2}{p}; \tag{3}$$

the second product on the right hand side of (3) is extended over *all* odd primes p under a certain (large, but not yet definitely characterized) upper bound. The first product is extended over the odd prime divisors of d; if d happens to be 2 (or a power of 2) there are no odd prime divisors, that first product is empty, and has to be replaced by 1. Therefore

$$P_d = \prod_{p|d} \frac{p-1}{p-2} \cdot P_2. \tag{4}$$

Yet the ratio of the probabilities P_d/P_2 should be approximately the same as the ratio of the observed numbers $\pi_d(x)/\pi_2(x)$—and so the formula (4) just derived justifies the conjectural formula (1)—complete success!

6. Unfortunately, our reasoning is vulnerable and the success is illusory. We left a gap in our derivation (we did not decide how far to go with p) and if we try to fill this gap, we run into trouble. The trouble becomes manifest if we try to apply our reasoning to the simplest analogous problem, the result of which is well known.

PROBLEM III. *Find the probability that x, a large integer chosen at random, is a prime number.*

By reasoning as we did in solving Problem I and assuming the independence of the probabilities involved as we did in solving Problem II we obtain the answer $\prod(p-1)/p$; the product is extended to all primes p not surpassing a certain bound—but what should be the bound? The number x is certainly prime if it is not divisible by any prime $p < x$. This leads to the evaluation of the desired probability

$$\prod_{p<x} \frac{p-1}{p} \sim \frac{\mu}{\log x} \tag{5}$$

where $\mu = 0.561459 \ldots = e^{-c}$ and $c = 0.577215 \ldots$ is the familiar constant of Mascheroni and Euler; the asymptotic evaluation in (5) (on the right hand side of the sign $\sim$) which is valid for $x \to \infty$, is due to Mertens.[6]

Now, the value (5) is too small. The probability in question is known to be $1/\log x$; this is just the prime number theorem. And we can "explain" somehow why the result is wrong: If the integer x is not divisible by any prime p which does not exceed $x^{1/2}$, x itself must be a prime—and so divisibility by primes exceeding $x^{1/2}$ is, in fact, *not* independent of the smaller primes.

Let us try to modify (5) by considering only primes p not exceeding $x^{1/2}$. This leads us to

$$\prod_{p\leqq x^{1/2}} \frac{p-1}{p} \sim \frac{\mu}{\log (x^{1/2})} = \frac{1.122\ldots}{\log x} \tag{6}$$

(we used Mertens' result (5)) and this value is too large.

Let us, however, imitate the physicists who, without hesitation, modify their theories to fit the observed facts. And so let us do a thing between (5) and (6) and extend the product to all *primes not exceeding* x^{μ}. We obtain so

$$\prod_{p<x^{\mu}} \frac{p-1}{p} \sim \frac{1}{\log x}, \tag{7}$$

the right result.

I do not pretend to understand why the introduction of the upper bound x^{μ} *should* yield the right result. For that matter, when the quanta were introduced, no physicist pretended to understand why energy should be obtainable (as salt or sugar is in the self-service store) only in uniform little packages, in multipla of a certain unit. Yet the criterion of a physical theory is its applicability. Let us apply the (unintelligible) trick that gave

[6] *Cf.* G. H. Hardy and E. M. Wright, *An Introduction to the Theory of Numbers*, Oxford, 1938, p. 349, Th. 430.

us the right expression for the prime number theorem to our formula (3). Extending the second product to *odd* primes p inferior to x^μ, we are led to

$$\tag{8} \begin{aligned} P_d &= \prod_{p|d} \frac{p-1}{p-2} \cdot \frac{1}{2} \prod_{p<x^\mu} \frac{p-2}{p} \\ &\sim \prod_{p|d} \frac{p-1}{p-2} \cdot 2 \prod_{p<x^\mu} \frac{(p-2)p}{(p-1)^2} \frac{1}{(\log x)^2}; \end{aligned}$$

we have used Mertens' result (5). It is easily seen that (8) is equivalent to

$$\tag{9} P_d \sim 2C_2 \prod_{p|d} \frac{p-1}{p-2} \frac{1}{(\log x)^2},$$

where C_2 stands for the convergent infinite product

$$\prod \left(1 - \frac{1}{(p-1)^2}\right)$$

extended to all odd primes $p = 3, 5, 7, 11, \ldots$. The asymptotic formula (9) is due to Hardy and Littlewood, yet even their argument, which is incomparably deeper and more difficult than the one presented here, does not prove (9); it just confers on (9) another kind of plausible evidence. Yet all available numerical data also seem to support (9).

Let us recall that we have attained (9) by combining two analogies, one of which was extremely "natural" and the other (the "trick of the magic μ") extremely "artificial." And let us try to draw the moral: mathematicians and physicists think alike; they are led, and sometimes misled, by the same patterns of plausible reasoning.[7]

---

[7] See G. H. Hardy and J. E. Littlewood, Some problems of "Partitio numerorium": On the expression of a number as a sum of primes, *Acta Mathematica*, vol. 44, 1922, pp. 1–70, especially Conjecture B on p. 42. The more general conjecture on p. 61 (Theorem X 1) is also obtainable by the foregoing reasoning. See also the literature quoted (and criticized) on pp. 32–34, especially the writings of Sylvester, concerning the use of probabilities in questions of similar nature. The crux of the matter may be so expressed: When we consider a fixed number of primes, the "probabilities" introduced can be regarded as "independent," but they cannot be so regarded when the number of primes considered increases in an arbitrary manner. [See also two papers in the *Quarterly Journal of Mathematics*, by Lord Cherwell, vol. 17, 1946, pp. 46–62, and by Lord Cherwell and E. M. Wright, 2. ser. vol. 11, 1960, pp. 60–63.]

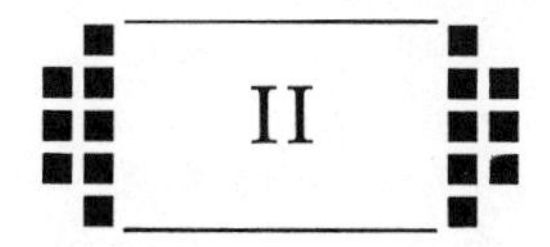

ADDITIONAL COMMENTS, PROBLEMS, AND SOLUTIONS

Each of the following comments refers to a section or example which is quoted at the beginning of the comment between square brackets [ ].

Each of the following problems fits into the collection of problems following a certain chapter at a definite place which is indicated by the numbering. Thus, the problem 1.10.1 should be considered as following ex. 1.10 and the problems 3.41.1, 3.41.2, 3.41.3, 3.41.4, and 3.41.5 as following ex. 3.41 in this order.

COMMENTS

[Sect. 3.18] The problem treated in sect. 3.8–3.18 and its solution are due to Jakob Steiner; see his *Gesammelte Werke*, vol. 1, pp. 77–94.

[Sect. 4.8] *Sums of squares and sums of factorials.* As we have seen in sect. 4.2, each of the first ten positive integers can be represented as the sum of not more than four squares. That the same is true for any positive integer has been conjectured by Bachet and proved later (first by Lagrange).

Let us pass from the squares

$$1, \quad 4, \quad 9, \quad 16, \ldots \quad n^2, \ldots$$

to the factorials

$$1, \quad 2, \quad 6, \quad 24, \ldots \quad n!, \ldots$$

and consider that

$$1 = 1!$$

$$2 = 2!$$

$$3 = 2! + 1!$$

$$4 = 2! + 2!$$

$$5 = 2! + 2! + 1!$$

$$6 = 3!$$

$$7 = 3! + 1!$$

$$8 = 3! + 2!$$
$$9 = 3! + 2! + 1!$$
$$10 = 3! + 2! + 2!$$

Should we conjecture on the basis of these facts that any positive integer can be represented as the sum of not more than three factorials?

Indeed, the statement "Each positive integer can be represented as the sum of not more than M factorials" is false for any M. In fact, as it is easy to see by mathematical induction,

$$(1) \quad n! - 1 = (n-1)!(n-1) + (n-2)!(n-2) + \dots + 2!2 + 1!1$$

for $n \geqq 2$ and, as it is not difficult to prove, $n! - 1$ cannot be represented as a sum of less factorials than

$$(n-1) + (n-2) + \dots + 2 + 1 = \frac{n(n-1)}{2}.$$

Proof. Let $a_1, a_2, \dots, a_{n-1}$ be non-negative integers such that

$$(2) \qquad (n-1)!a_{n-1} + (n-2)!a_{n-2} + \dots + 2!a_2 + 1!a_1 = n! - 1$$

and $a_1 + a_2 + a_3 + \dots + a_{n-1}$ is a *minimum;* $n > 2$.

In the first place,

$$a_{n-1} \leqq n - 1;$$

otherwise the left-hand side of (2) would be too large. Then

$$a_{n-2} \leqq n - 2;$$

otherwise we could replace a_{n-2} by $a_{n-2} - (n-1)$ and a_{n-1} by $a_{n-1} + 1$ and thereby decrease the sum $a_1 + a_2 + \dots + a_{n-1}$. Similarly

$$a_{n-3} \leqq n - 3,$$
$$\dots\dots\dots$$
$$a_2 \leqq 2,$$
$$a_1 \leqq 1.$$

Yet, in all the $n - 1$ inequalities just proved, the case of equality must hold: otherwise the left-hand side of (2) would be too small, see (1).

And so we conclude that the minimum of $a_1 + a_2 + \dots + a_{n-1}$ is indeed $1 + 2 + \dots + (n-1)$.

[Ex. 6.9] In a paper of the author (On picture-writing, *American Mathematical Monthly*, vol. 63, 1956, pp. 689–697) this result and two similar ones are derived by graphic considerations which are appropriate to render the concept of generating functions intuitive. The problem treated in ex. 6.7–

6.9 is due to A. de Segner; see Euler, *Opera Omnia*, ser. 1, vol. 26, pp. XVI–XVIII.

[Sect. 8.3] *More on the pattern of the tangent level line: A "dual" pattern.*

(1) *Example.* Given two intersecting straight lines, a and b, and a point C inside the angle formed by a and b. Draw the straight line x through C that makes the perimeter of the triangle formed by a, b, and x a minimum.

As hinted by Fig. A1, we consider only triangles contained in that one of the four angles formed by a and b that contains the given point C.

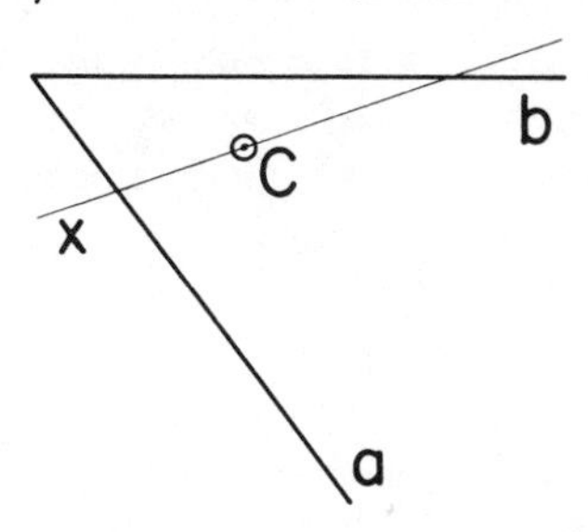

Fig. A1. Looking for the shortest perimeter.

The proposed problem is not too easy. Let us begin by explaining a useful lemma.

(2) *A locus for a variable straight line.* Fig. A2 shows a circle tangent to both sides of $\angle AVB$ at the points A and B, respectively; the line EF touches

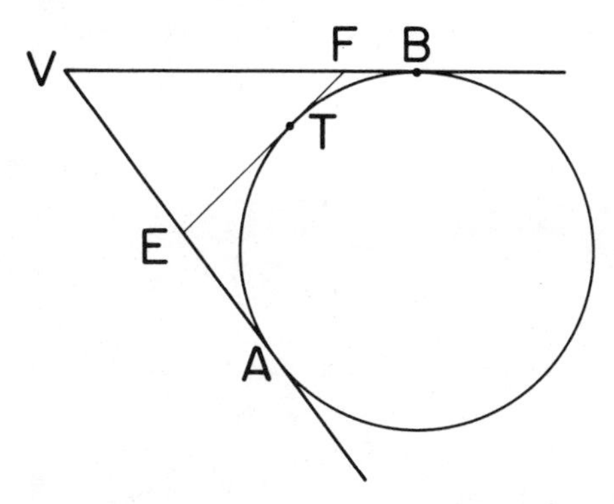

Fig. A2. A locus for a straight line, also a kind of level line.

the circle at the point T. The tangents to a circle from an external point are equal and so

$$ET = EA, \qquad FT = FB.$$

Hence

$$VE + ET = VE + EA = VA$$

$$VF + FT = VF + FB = VB$$

and therefore

$$VE + VF + EF = VA + VB = 2VA.$$

Observe that the length VA is *independent* of the situation of the point T on the circular arc AB. After this observation only a little additional consideration is needed to prove the proposition: *The set of all straight lines that cut off a triangle of given perimeter from a given angle consists of the tangents to an arc of circle touching the given angle at its endpoints and turning its convex side to the vertex of the angle.*

This arc (ATB in Fig. A2) is properly called in this connection a "locus," not of its points, but of its tangents.[1] The same arc can also be considered as a kind of level line: We consider the perimeter of the triangle cut off from the given angle as a function of the cutting straight line; this function remains constant when the cutting straight line ranges over all positions tangent to the arc.

(3) Now we are very close to the solution of the proposed problem, the idea may easily appear. Let us draw several circles touching the given lines a and b. If such a circle has two tangents passing through C, as x and x' in Fig. A3, these two tangents cut off two triangles of equal perimeter

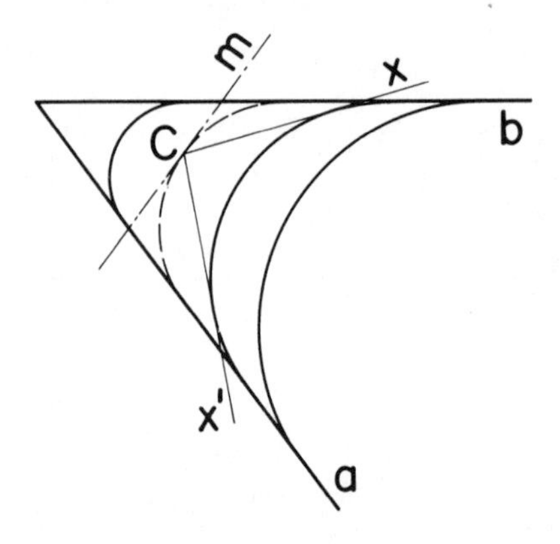

Fig. A3. When the two tangents through the prescribed point coincide.

from the angle, but this perimeter is not the shortest possible: an arc nearer to the point C yields a shorter perimeter. Circles with two tangents through C cannot do the trick: the circle touching a and b and *passing through* C has a unique tangent at C (m in Fig. A3) and *this* tangent cuts off the triangle with minimum perimeter.

(4) The reader should compare the foregoing paragraph sentence by sentence with the last paragraph of sect. 8.2 and compare also Fig. A3 with Fig. 8.3. Afterwards he should try to formulate the general pattern under-

[1] *Cf.* the author's *Mathematical Discovery*, vol. 1, p. 147, ex. 6.8 and its solution on p. 204.

lying the foregoing solution; he can obtain it from the pattern of sect. 8.3 by "duality," that is, by interchanging consistently the terms "point" and "line" (that is, straight line).

It may be useful to adapt our terminology to the situation. First, let us say that two curves tangent to each other "have two coinciding points of intersection." Second, let us call the set of (straight) lines passing through the given point C (of our problem just solved) the "prescribed set." Then we may express the pattern introduced in sect. 8.3 so:

The level line that yields the extremum has two coinciding points of intersection with the prescribed path.

Now, the following description applies to the foregoing solution and, we hope, also to analogous cases:

The level line that yields the extremum has two coinciding tangents belonging to the prescribed set.

[Ex. 10.41] *The principle of non-sufficient reason.* Readers familiar with the philosophy of Leibnitz may be reminded, by reading ex. 10.41, of his celebrated "Principle of Sufficient Reason," a certain version of which was given the provocative name "Principle of Non-Sufficient Reason." The Principle of Sufficient Reason is often mentioned in Leibnitz's works, but I know only of one passage, found in his manuscripts and posthumously printed, that explicitly refers to mathematical matters;[2] it adds to the piquancy of this passage that one word of it was deleted, and another added later to it, by Leibnitz. Perhaps Leibnitz hesitated between two statements which, in more modern terms and sufficiently clarified, may be formulated as follows:

I. *Unknowns with respect to which the condition is symmetric must obtain the same value in the solution,* IF *the solution is unique.*

II. *Unknowns with respect to which the condition is symmetric may be expected to obtain the same value in the solution.*

Statement I has been essentially proved in ex. 10.41, but statement II is purely heuristic. For further explanations and especially for illustrative examples I must refer to my other writings.[3]

[Ex. 10.43] There is a polyhedron with 20 faces that has an isoperimetric quotient superior to that of the regular icosahedron; *cf.* Goldberg, *loc. cit.* ex. 10.42.

[Sect. 11.4] *An argument of Galileo.* We used a differential equation in discussing Galileo's rejection of the law of free fall that he had originally assumed. Galileo could not use calculus which had not yet been invented in his time, but he succeeded nevertheless in devising a cogent mathematical argument, which is stated in his "Discorsi e Dimostrazioni" very clearly, but

---

[2] Louis Couturat, *Opuscules et fragments inédits de Leibniz*, 1903; see p. 519.

[3] *Mathematical Discovery*, vol. 2 ex. 15.21–15.40, and "Inequalities and the Principle of Non-Sufficient Reason," *Inequalities*, ed. Oved Shisha, New York, London, 1967, pp. 1–15.

perhaps too concisely and without a figure.[4] This conciseness may have led even such an authority as Mach to misjudge Galileo's argument.[5] And so it may be worth while to restate Galileo's remarkable argument with more detail and in more modern language.

The assumption that will be refuted is stated by Simplicio, the refutation is explained by Salviati. This is the usual distribution of roles in Galileo's great dialogues: Simplicio states the wrong opinion, Salviati represents Galileo's own standpoint.

SIMPLICIO. I believe that the velocity of a falling body starting from rest is proportional to the distance traveled, so that the velocity is doubled when the body falls from a doubled height. It seems to me that this proposition is so obvious that it should be accepted without hesitation and contradiction.

SALVIATI. And yet it is as false and impossible as instantaneous, timeless motion. Listen to the following argument. Consider two bodies, starting from rest, from the same height, and at the same instant. Both move according to the same law that seems obvious to Simplicio: their velocities bear the same proportion to the distance traversed. The first body starts from the point a, passes successively through the points b, c, and d and attains the endpoint e ten yards below a. The second body starts from the point A, and after passing the points B, C, and D attains E twenty yards below A; see Fig. A4. The notation is so chosen that the points correspond:

$$a \text{ to } A, \qquad b \text{ to } B, \qquad c \text{ to } C, \qquad d \text{ to } D, \qquad e \text{ to } E,$$

and any distance in the second motion is double the corresponding distance in the first:

$$AB = 2ab, \qquad AC = 2ac, \qquad AD = 2ad, \qquad AE = 2ae.$$

By the law assumed by Simplicio, the velocity at any point of the second motion (such as B in Fig. A4) is double the velocity at the corresponding point of the first motion (which is b in our case—since $AB = 2ab$). Thus in passing from the first motion to the second, we see that both the distances and the velocities are doubled—yet then the time required remains *unchanged* so that corresponding points are attained at the same time, for example e the same instant as E.[6] In fact, however, a body falling according to the

---

[4] *Le Opere di Galileo Galilei*, edizione nazionale, vol. 8, pp. 203–204. For a translation, see Galileo Galilei, *Dialogues concerning two new sciences*, Dover, pp. 167–168.

[5] Ernst Mach, *Die Mechanik*, 7th edition, 1912, p. 122: "Er lehnt nun die erstere Auffassung aus ebenso unzutreffenden Gründen ab, als er sie früher angenommen hatte."

[6] This conclusion can be made intuitive and, in fact, derived without the notation of the integral calculus with which, however, the derivation becomes very brief (t is time, v velocity, x distance):

$$t = \int \frac{dx}{v} = \int \frac{d(2x)}{2v}.$$

assumed law would perform BOTH motions (the second being an extension of the first—both obey the SAME law!) and so it would require no time for passing from the elevation e to the elevation E. Such an instantaneous, timeless motion is contrary to all observation—the assumption led us to an absurdity.

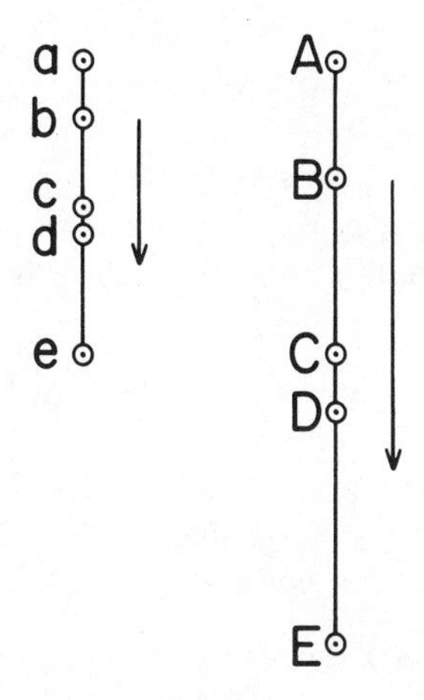

Fig. A4. Velocity proportional to distance traversed.

[Ex. 13.11] My paper "On patterns of Plausible Inference," *Studies and Essays presented to R. Courant*, 1948, pp. 277–288, adds one more illustration to those given in ex. 13.12–13.19.

[Ex. 13.15] Has the fact that all roots of the equation that "sets the fashion" coincide some connection with the Principle of Non-Sufficient Reason? *Cf.* the comment in this Appendix on ex. 10.41.

[Ex. 13.20] As this Appendix goes to press, the following changes must be recorded: The conjecture mentioned in ex. 13.16 has been refuted by C. B. Haselgrove (*Mathematika*, vol. 5, 1958, 141–145) who followed up A. E. Ingham's remarks and used a computing machine. The conjecture mentioned in ex. 13.17 has been proved by M. Schiffer (*Comptes Rendus*, Paris, vol. 244, 1957, pp. 3118–3121). The conjecture advanced in ex. 13.18 remains unproved and unrefuted, yet an extensive particular case of it has been verified (see the author's paper, *Proceedings London Math. Soc.* (3), vol. 11, 1961, pp. 419–433).

[Sect. 15.3] The last sentence should be expanded as follows: You can build a machine to draw demonstrative conclusions for you but, I think, you can never build a machine that will draw plausible inferences *all by itself*. It could prepare, however, the materials from which *you* may draw the desired plausible inference.

[Sect. 16.9] It is appropriate to quote here a paper by the author in which heuristic considerations take up somewhat more space than usual; see *Mathematische Annalen*, vol. 114, 1937, pp. 622–634, especially p. 623, footnote 2.

PROBLEMS

1.10.1. (1) Assuming the truth of Goldbach's conjecture, derive the following theorem: *Let n be an integer* $n \geqq 3$. *Then there exists an odd prime number p such that* $n \leqq p \leqq 2n - 3$.

(2) Indeed, the theorem just stated has been proved, independently of any assumption, by the great Russian mathematician Tchebyshev. How does this fact influence your judgment? Does it render Goldbach's conjecture more or less likely?

3.8.1. The side of an equilateral triangle is of length n (n is an integer). By straight lines parallel to its sides, the triangle is subdivided into equilateral triangles with sides of length 1.

(1) Express N_0, N_1, and N_2 (defined in ex. 3.7) in terms of n.

(2) Is the relation ex. 3.7 (2) valid in the present case?

3.40.1. Any two vertices of a certain convex polyhedron are connected by an edge of the polyhedron. Show that this polyhedron is a tetrahedron.

3.40.2. (1) Prove that

$$(E + 6)/3 \leqq F \leqq 2E/3.$$

(2) Try to find a convex polyhedron for which $E = 7$.

3.41.1. Prove that

$$F_3 + V_3 \geqq 8.$$

3.41.2. *The "converse" of Euler's theorem.* Prove the proposition: *Three integers*

$$F, \quad V, \quad \text{and } E$$

represent the number of faces, vertices, and edges of some convex polyhedron, respectively, if, and only if, they satisfy the equation

$$F + V = E + 2$$

and the two inequalities

$$2E \geqq 3F, \qquad 2E \geqq 3V$$

3.41.3. Prove that numbers F, V, and E satisfying the system of one equation and two inequalities considered in ex. 3.41.2 are necessarily positive.

3.41.4. Prove that the system of one equation and two inequalities considered in ex. 3.41.2 is equivalent to the system consisting of the same equation and of the two following inequalities:

$$2V - 4 \geqq F, \qquad 2F - 4 \geqq V.$$

3.41.5. Show that from any convex polyhedron with

V vertices, F faces, and E edges

of which not all the faces are triangles it is possible to derive another convex polyhedron with

V vertices, $F + 1$ faces, and $E + 1$ edges.

4.15.1. (1) $R_4(n)/8$ is odd in just two cases: first, when n is a square, and, secondly, when $n/2$ is a square. Prove this statement and show that the conjecture arrived at in the solution of ex. 4.14 agrees with it.

(2) How does the remark (1) influence your judgment? Does it render that conjecture more or less likely?

7.11.1. "A convex polygon with n sides is dissected into $n - 2$ triangles by $n - 3$ diagonals."

Prove this statement (assumed as obvious in ex. 6.7) by mathematical induction.

8.18.1. Of a tetrahedron, given L, the sum of the lengths of three edges which do *not* surround a face (two different configurations are possible). Find the maximum of the volume of the tetrahedron.

8.18.2. Of a tetrahedron, given L, the sum of the lengths of four edges which form a skew quadrilateral. Find the maximum of the volume of the tetrahedron.

8.18.3. Of a box, given L, the sum of the lengths of seven edges which form an open skew polygonal line in joining the eight vertices of the box, each one to the next, in the following sequence (the points are characterized by their rectangular coordinates):

$(0,0,0)$ $(a,0,0)$ $(a,b,0)$ $(0,b,0)$ $(0,b,c)$ (a,b,c) $(a,0,c)$ $(0,0,c)$.

Find the maximum of the volume of the box.

(Compare with ex. 8.18.1 and try to think of a more general problem.)

8.18.4. Of a box, given L, the sum of the lengths of eight edges which form a skew octagon: to the seven edges listed in ex. 8.18.3 add the one that joins the vertex $(0,0,c)$ to $(0,0,0)$. Find the maximum of the volume of the box.

(Compare with ex. 8.18.2 and try to think of a more general problem.)

8.43.1. You could repeat the considerations of ex. 8.43 in three, instead of two, dimensions. Use the simplest case to prove that

$$(u_1v_1 + u_2v_2 + u_3v_3)^2 \leqq (u_1^2 + u_2^2 + u_3^2)(v_1^2 + v_2^2 + v_3^2)$$

where equality holds if, and only if,

$$u_1 : v_1 = u_2 : v_2 = u_3 : v_3.$$

8.58.1. Compare the two following problems:

Given the perimeter L of a sector of a circle. Find the shape of the sector for which its area A is a maximum.

Given the perimeter L of a rectangle. Find the shape of the rectangle for which its area A is a maximum.

These problems are obviously analogous. Make the analogy as perfect as possible.

8.63.1. Let Λ denote the principal frequency of a homogeneous rectangular membrane with sides a and b; then

$$\Lambda^2 = \pi^2 \left(\frac{1}{a^2} + \frac{1}{b^2}\right)$$

(except for a factor of proportionality which depends on the material, thickness, and tension of the membrane). Show that of all rectangular membranes with a given area A the square membrane has the lowest principal frequency.

8.63.2. Let I denote the polar moment of inertia of a homogeneous rectangular plate with respect to its center of gravity; then

$$I = ab(a^2 + b^2)/12$$

(except for a certain "physical" factor of proportionality); a and b are the sides of the rectangle. Show that of all rectangular plates with a given area A the square plate has the minimum polar moment of inertia.

8.63.3. Let I denote the polar moment of inertia of a homogeneous triangular plate with respect to its center of gravity; then

$$I = A\frac{a^2 + b^2 + c^2}{36}$$

(except for a certain "physical" factor of proportionality, as in ex. 8.63.2); a, b, and c are the lengths of the sides of the triangle, A its area. Show that of all triangular plates with a given area the equilateral plate has the minimum polar moment of inertia.

8.63.4. Let a, b, and c denote the lengths of the sides and A the area of a triangle. Show that

$$A \leqq \frac{a^2 + b^2 + c^2}{4\sqrt{3}}$$

with equality for the equilateral, and no other, triangle.

(Ex. 8.40, ex. 8.43.1.)

9.28.1. *The golden mountain* of the fable has been recently discovered, according to a so-called reliable source. There is, we are told, a whole island that consists of solid gold.

According to the same source, you will receive, for your well-known merits, a piece of area A of this island. More precisely, A is the area of the horizontal projection of that piece of land the location of which you can still choose.

You must know that you will be allowed to exploit the solid golden rock beneath your piece of land down to sea level, but not farther.

I foresee that you will choose a piece of land around the top of the mountain. Describe its location in precise terms and prove that your choice makes the maximum volume of gold available.

10.26.1. *A more general post office problem.* (*Cf.* ex. 8.62.) We consider a right cylinder and let V, h, A, and L denote its volume, its height, the area of its base, and the perimeter of its base, respectively. We do not know the shape of the base (it may be surrounded by a closed curve of arbitrary form) but we do know that the cylinder's height and girth combined do not exceed l inches ($h + L \leqq l$). Given l, find the maximum of the volume V.

10.26.2. *The least convex solid containing a given arc of curve in space* is called the "convex hull" of that arc. Let L denote the length of the arc and V the volume of its convex hull. Prove that

$$V \leqq L^3/(18\pi\sqrt{3})$$

and that the equality is attained only if the arc is one full turn of a helix with pitch $1/\sqrt{2}$.

(*Cf.* ex. 8.18.1, 8.18.3. In seeking a proof you may consider simplifying assumptions about the nature of the arc examined.)

10.26.3. (continued). In order to conceive more clearly the solid figure, consider its orthogonal projection onto a plane. Which plane has the best chance to be useful?

10.26.4. Find the volume of a tetrahedron being given c, the length of one of its edges, and P, the area of its orthogonal projection onto a plane perpendicular to that edge.

10.40.1. Of all plates with a given area, the circular plate has the minimum polar moment of inertia.

Prove this assertion, advanced in ex. 10.40, which is much more accessible than the other similar assertions discussed there. (Superpose an arbitrary figure and a circle, both of the same area, so that their centers of gravity coincide.)

10.43.1. *Space analogue to ex.* 10.5. Prove that a sphere has a larger volume than any circumscribable polyhedron with the same surface area.

12.3.1. Keep on considering the triangle treated in ex. 12.3 and let r denote the radius of the circle circumscribed about it. It is asserted that

$$4Ar = abc.$$

Check this assertion in as many ways as you can.

12.5.1. Keep on considering the tetrahedron treated in ex. 12.5, let R denote the radius of the sphere circumscribed about it, and put

$$P = (ae + bf + cg)/2.$$

It is asserted that

$$36V^2R^2 = P(P - ae)(P - bf)(P - cg).$$

Check this assertion (which shows remarkable analogy to Heron's formula) in as many ways as you can.

12.5.2. Consider a spherical triangle on the unit sphere. Its sides are a, b, c, its angles A, B, C; each of these six quantities is positive and less than π. Let V denote the volume of the tetrahedron one vertex of which is the center of the sphere and the other three vertices the vertices of the spherical triangle. Set

$$a + b + c = 2s, \qquad A + B + C = 2S.$$

It is asserted that

$$9V^2 = \sin s \sin (s - a) \sin (s - b) \sin (s - c),$$

$$V = -\frac{2}{3} \frac{\cos S \cos (S - A) \cos (S - B) \cos (S - C)}{\sin A \sin B \sin C}.$$

Notice the analogy of these expressions to Heron's formula and check them in as many ways as you can.

16.6.1. "In any triangle the sum of the three . . . is greater than the semiperimeter."

Replace the dots . . . successively by

(I) altitudes

(II) medians

(III) bisectors (of the angles).

You obtain so three different assertions. Examine each assertion: is it true or false? Prove your answer!

16.6.2. If the area of a triangle is rational (that is, measured by a rational number) there are four thinkable cases: The triangle may have three or two rational sides, or just one or no rational side. Show by (preferably simple) examples that all four cases are actually possible.

SOLUTIONS

1.10.1. (1) Assume that

$$2n = p + p', \qquad p \text{ and } p' \text{ primes}, \qquad p \geqq p' \geqq 3.$$

It follows $n \leqq p \leqq 2n - 3$.

(2) More likely, most people would say: the confirmation of a consequence strengthens the conjecture.

3.8.1.

(1) $N_0 = 1 + 2 + 3 + \ldots + (n + 1)$,
$N_1 = 3(1 + 2 + 3 + \ldots + n)$,
$N_2 = n^2$.

(2) $$\frac{(n+1)(n+2)}{2} - \frac{3n(n+1)}{2} + n^2 = 1.$$

3.40.1. The condition of the problem yields the first part and ex. 3.40 the second part of the double inequality:

$$V(V-1)/2 \leqq E \leqq 3V - 6$$

from which follows

$$V^2 - 7V + 12 = (V-3)(V-4) \leqq 0.$$

The only integers satisfying this condition are $V = 3$ and $V = 4$. Yet $V = 3$ is obviously excluded, and the tetrahedron is the only polyhedron with four vertices.

3.40.2. (1) Ex. 3.40. (2) Part (1) yields for $E = 7$

$$4\tfrac{1}{3} \leqq F \leqq 4\tfrac{2}{3}$$

and so F cannot be an integer: there is no convex polyhedron with seven edges.

3.41.1. $4F - 2E + 4V - 2E = 8$, ex. 3.31, and ex. 3.32:

$$F_3 - F_5 - 2F_6 - 3F_7 - \ldots + V_3 - V_5 - 2V_6 - 3V_7 - \ldots = 8.$$

3.41.2. In view of the foregoing, and of the following ex. 3.41.3 and 3.41.4, it remains the task: Given a positive integer V greater than 3, for each positive integer F that lies between the limits indicated by

$$\text{(I)} \quad (V + 4)/2 \leqq F \leqq 2(V - 2)$$

exhibit a convex polyhedron with

V vertices and F faces

and, of course, with $V + F - 2$ edges.

(1) Let us point out the existence of polyhedra exhibiting some of the simplest combinations.

$V = 4$, $F = 4$: tetrahedron.

$V = 5$, $F = 5$: pyramid with quadrilateral base.

$V = 7$, $F = 6$: obtained from the foregoing pyramid by "truncating," that is, cutting off a three-edged vertex (a vertex of the base; *cf.* sect. 3.4).

(2) $V = 2n$ even, $n \geqq 3$, and F has the minimum value compatible with (I), that is,

$$F = (2n + 4)/2 = n + 2.$$

The prism of which the base has n sides exhibits this combination of V and F.

(3) $V = 2n + 1$ odd, $n \geqq 4$, and F has the minimum value compatible with (I), that is,

$$F = \{(2n + 1 + 4)/2\} + \tfrac{1}{2} = n + 3.$$

Start from a prism the base of which has $n - 1$ sides (observe that $n - 1 \geqq 3$) and consider two of its vertices, one belonging to the base, the other to the upper base, joined by a lateral edge l of the prism. (See Fig. A5, left-hand side.) Cut off both vertices so that the intersection of the

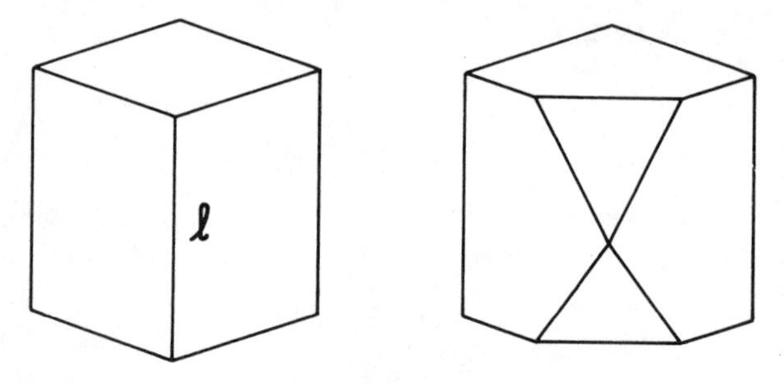

Fig. A5. Cutting off two corners of a prism.

two cutting planes has a common point with l—at which common point a four-edged vertex of the new polyhedron will be formed. For this new polyhedron (see Fig. A5, right-hand side)

$$V = 2n + 1,$$

$$F_3 = 2, \qquad F_4 = n - 3, \qquad F_5 = 2, \qquad F_n = 2, \qquad F = n + 3,$$

as required. (The evaluation for F is correct for $n \geqq 4$, but for F_k with $k \leqq 5$ only for $n \geqq 6$.)

(4) As (1), (2), and (3) show, there exists, for $V = 4,5,6, \ldots$, a polyhedron having F faces where F is the least possible integer satisfying (I).

(5) Take now any integer F that satisfies (I) without being the greatest possible number satisfying it and about which we know already that there is a polyhedron having V vertices and F faces. This polyhedron has at

least one face which is not a triangle, by ex. 3.41.4, since $F < 2V - 4$. Hence, by ex. 3.41.5 (Fig. A6), there exists also a polyhedron with V vertices and $F + 1$ faces.

In view of (4), we can exhaust so step by step all combinations of V and F compatible with (I) and fully prove our assertion.

3.41.3.
(1) $3E + 6 = 3F + 3V \leqq 4E$, hence $E \geqq 6$.
(2) $3F = E + (2E - 3V) + 6 \geqq 6 + 0 + 6$, hence $F \geqq 4$.
(3) Similarly, $V \geqq 4$; *cf.* ex. 3.4.

3.41.4. From the equation and the first inequality there follows

$$2F + 2V = 2E + 4 \geqq 3F + 4, \text{ hence } 2V - 4 \geqq F$$

and this derivation is reversible: from the inequality just derived and the equation follows $2E \geqq 3F$. Observe that, by ex. 3.40, in both inequalities considered the case of equality is attained if, and only if, all faces are triangular.

The remaining half of the assertion is proved in the same way.

3.41.5. Take a non-triangular face of the polyhedron with V, F, E, draw in it a diagonal (dotted in Fig. A6) dividing the face in two parts, tilt "slightly

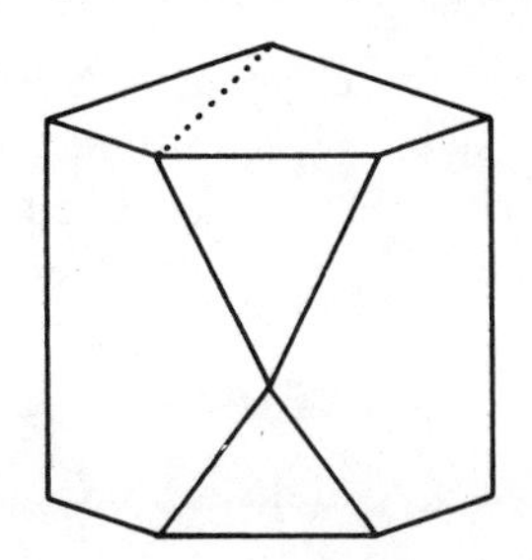

Fig. A6. From F to $F + 1$.

inward" one of the parts, rotating it about that diagonal, and obtain so a new polyhedron with V, $F + 1$, $E + 1$.

4.15.1. (1) The table displayed in the solution of ex. 4.9 contains only two cases, (10) and (11), in which the co-factor of 8 is odd.

To show that the conjectured number behaves the same way, let u stand for an odd integer and distinguish three cases:

(I) $n = u$,

(II) $n = 2u$,

(III) $n = 2^{\alpha}u$ with $\alpha > 1$.

To treat the cases (I) and (II), use the solution of ex. 4.8 and reduce the case (III) to (I) and (II).

(2) More likely, most people would say.

7.11.1. The statement is trivially valid for $n = 3$. Assume that it is valid for $3,4,5, \ldots n-1$ where $n > 3$, and draw a diagonal dividing the polygon with n sides into two subpolygons, one with k sides, the other with l sides. Then

$$k + l = n + 2, \qquad k \leqq n - 1, \qquad l \leqq n - 1$$

and so the assertion is, by our assumption, valid for both subpolygons: they are divided into

$$k - 2 \text{ triangles by } k - 3 \text{ diagonals,}$$

$$l - 2 \text{ triangles by } l - 3 \text{ diagonals,}$$

respectively. Hence, the n-sided polygon itself is divided into

$$k - 2 + l - 2 = n - 2 \text{ triangles by}$$

$$k - 3 + l - 3 + 1 = n - 3 \text{ diagonals.}$$

Thus we have proved the desired statement for n.

8.18.1. The three edges may either start from the same vertex as in ex. 8.18 or they may form an open polygonal line.

(1) Let V denote the volume. If the lengths a, b, and c of the three edges are given individually (as in ex. 8.18) we have (as there) for both configurations:

$$V \leqq abc/6.$$

(2) If just L is given, but a, b, and c are not,

$$a + b + c = L$$

and we conclude from the theorem of the means (sect. 8.6) that

$$V \leqq abc/6 \leqq L^3/162.$$

8.18.2. In the tetrahedron, let us call *linked edges* the four sides of the quadrilateral considered and *free edges* the remaining two (which are, in fact, opposite edges of the tetrahedron). By partial variation, we find that the dihedral angles at both free edges are right angles and that all four linked edges are of the same length $L/4$.

Hence the four faces of the tetrahedron are congruent isosceles triangles, each has two legs of length $L/4$; let b denote the length of the base (a free edge) and h that of the height perpendicular to the base; let V stand for the volume of the tetrahedron. Then (take into account what we have said about the dihedral angles at the free edges)

$$h^2 + b^2/4 = L^2/16$$

$$b^2 = 2h^2$$

$$V = 2(h^2/2)(b/2)/3 = L^3/(288\sqrt{3})$$

8.18.3. Obviously

$$L = 4a + 2b + c$$

$$V = abc = 4a \cdot 2b \cdot c/8$$

$$\leqq (L/3)^3/8 = L^3/216;$$

we have used the theorem of the means, see sect. 8.6. The case of equality, and so the maximum of V, is attained when

$$4a = 2b = c = L/3.$$

Ex. 8.18.1 and 8.18.3 suggest ex. 10.26.2.

8.18.4. Now

$$L = 4a + 2b + 2c$$

$$V = abc = 4a \cdot 2b \cdot 2c/16$$

$$\leqq (L/3)^3/16 = L^3/432$$

with equality for

$$2a = b = c = L/6.$$

Ex. 8.18.2 and 8.18.4 suggest to change ex. 10.26.2 by considering a closed space curve instead of an open arc of such a curve. See the end of the solution of ex. 10.26.2.

8.43.1. Consider the triangle whose vertices are

$$(0,0,0), \qquad (u_1,u_2,u_3), \qquad (-v_1,-v_2,-v_3).$$

The sum of the lengths of two sides is greater than the length of the third side and so

$$[(u_1 + v_1)^2 + (u_2 + v_2)^2 + (u_3 + v_3)^2]^{1/2} < [u_1^2 + u_3^2 + u_3^2]^{1/2} + [v_1^2 + v_2^2 + v_3^2]^{1/2}$$

unless the triangle degenerates and the proportion given in the problem holds. Square both sides and simplify.

The passage from two to three dimensions suggests a further passage from three to n.

8.58.1. Let r stand for the radius and s for the arc of the sector. Then

$$A = r\frac{s}{2}, \qquad L = 2\left(r + \frac{s}{2}\right).$$

Let a and b stand for two sides of the rectangle starting from the same vertex. Then

$$A = ab, \qquad L = 2(a + b).$$

If we set

$$r = a, \qquad \frac{s}{2} = b$$

the two problems become equivalent, the solution of one yields immediately the solution of the other, and so, by sect. 8.5, in both cases

$$A \leqq L^2/16.$$

8.63.1.

$$\frac{A\Lambda^2}{2\pi^2} = \frac{a^2 + b^2}{2ab} \geqq 1$$

and equality is attained if, and only if, $a = b$.

8.63.2.

$$\frac{6I}{A^2} = \frac{a^2 + b^2}{2ab} \geqq 1$$

and equality is attained if, and only if, $a = b$.

8.63.3. By ex. 8.63.4

$$I \geqq A^2/(3\sqrt{3})$$

with equality for the equilateral triangle only.

8.63.4. Heron's formula, see ex. 8.40, can be transformed into

$$16A^2 = (a^2 + b^2 + c^2)^2 - 2(a^4 + b^4 + c^4).$$

By ex. 8.43.1

$$3(a^4 + b^4 + c^4) \geqq (a^2 + b^2 + c^2)^2$$

with equality for $a^2 = b^2 = c^2$. Combining the two preceding results we obtain

$$16A^2 \leqq (a^2 + b^2 + c^2)^2/3.$$

9.28.1. (Some readers may prefer to restate the problem in the terminology of the integral calculus: Given the function $f(x,y)$ and the positive number A, the area of the domain of integration of the double integral

$$\iint f(x,y)\,dxdy$$

find the shape and location of that domain of integration for which the value of this double integral becomes a maximum.) The desired maximum is attained when the boundary of the piece of land (whose horizontal projection is the domain of integration) is a level line of the golden mountain (of $f(x,y)$) throughout the interior of which the elevation (that is, $f(x,y)$) is higher than anywhere outside.

For a proof, compare the horizontal projection of the interior R of the level line described, represented as a circle in Fig. A7, with any other region R' of the same area A and in the same plane, represented as a rectangle in

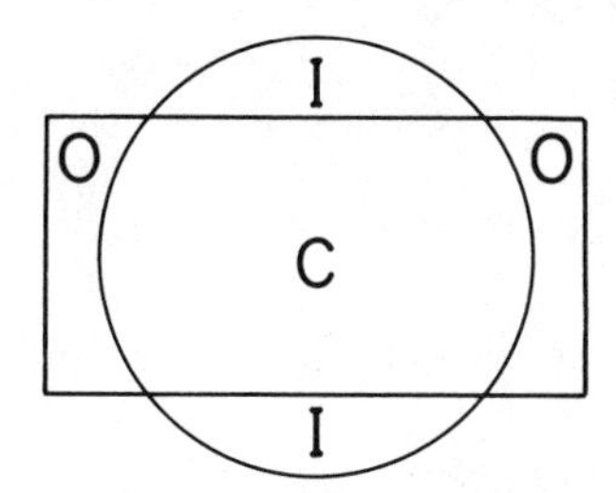

Fig. A7. The level line yields the extremum.

Fig. A7. Let C denote the common part of the two regions R and R', and set symbolically

$$R = C + I,$$

$$R' = C + O.$$

That is, R consists of the two non-overlapping subregions C and I, and R' of C and O; I is inside, O is outside the projection of the important level line. The difference of the two volumes over R and R' is the same as that of the two volumes over I and O. Yet this latter difference is obviously positive, because the two volumes have projections of equal area (A less the area of C), and the elevation in any inner point of I is superior to that in any inner point of O. (Repeat this proof in the notation of the integral calculus if you are familiar with it.)

10.26.1. By the isoperimetric inequality (see sect. 10.8) we obtain

$$V = Ah \leqq L^2h/(4\pi). \tag{1}$$

By the theorem of the means (see sect. 8.6)

$$\left(\frac{L}{2}\right)^2 h \leqq \left(\frac{\frac{L}{2} + \frac{L}{2} + h}{3}\right)^3 \leqq \left(\frac{l}{3}\right)^3. \tag{2}$$

In combining (1) and (2) we obtain

$$V \leqq l^3/(27\pi).$$

The right-hand side is the desired maximum; it cannot be attained unless the base of the cylinder is a circle, see (1), with a radius r such that

$$\pi r = L/2 = h = l/3,$$

see (2).

10.26.2. Let us call the arc considered C, its convex hull H, and the straight line segment joining the two endpoints of C the *principal chord* of C; *cf.* ex. 10.26.3. A triangle of which the base is the principal chord and the vertex opposite the base a point of C, belongs obviously to H; *assume*, in restricting the nature of C, that H consists entirely of such triangles.

We choose a plane perpendicular to the principal chord and we call it the *principal plane;* cf. ex. 10.26.3. The orthogonal projection of C onto the principal plane is a closed curve C'. One point of C' is the projection of the two endpoints of C; *assume* that any other point of C' is the projection of just one point of C.

(Look at the two particular cases considered in ex. 8.18.1 and 8.18.3. Which assumptions are, and which are not, fulfilled by each?)

Let

c denote the length of the principal chord,

L' the length of C',

A' the area enclosed by C'.

By half planes whose edge contains the principal chord we divide H into "approximate tetrahedra" and applying ex. 10.26.4 we obtain

$$V = A'c/3. \tag{1}$$

By the isoperimetric inequality (see sect. 10.8) we obtain

$$A' \leqq L'^2/(4\pi). \tag{2}$$

Now, let the principal plane be horizontal. Consider the infinite cylindrical surface with vertical generating lines the orthogonal cross-section of which is C'. Unroll this cylindrical surface, which also contains C, into a plane. By such unrolling C' goes over into a straight line segment of length L', one leg of a right triangle of which the other leg is the principal chord of length c, and C goes over into an arc in the plane joining the two endpoints of the hypotenuse of that right triangle. Yet the shortest way between two points is the straight line, and so

$$L'^2 + c^2 \leqq L^2. \tag{3}$$

By combining (1), (2), and (3) we derive that

$$V \leqq \frac{c(L^2 - c^2)}{12\pi}. \tag{4}$$

By the theorem of the mean (see sect. 8.6)

$$2c^2(L^2 - c^2)(L^2 - c^2) \leqq (2L^2/3)^3. \tag{5}$$

By combining (4) and (5) we obtain the desired result.

The case of equality cannot be attained unless C' is a circle, C becomes a straight line by unrolling, and

$$2c^2 = L^2 - c^2;$$

examine (2), (3), and (5), respectively.

The result obtained is due to E. Egerváry[7] who derived both assumptions made here from one geometrically simple assumption. Whether the result holds without any such restrictive assumptions remains an open question.

The corresponding question concerning closed curves was treated by I. J. Schoenberg in 4,6,8, . . . dimensions.[8]

10.26.3. The chord joining the two endpoints of the arc considered is the only one of its kind, and a plane perpendicular to this chord also has a privileged position in relation to the arc. Try to project the space figure onto such a plane.

10.26.4. $Pc/3$.

10.40.1. Repeat, with appropriate changes, the solution of ex. 9.28.1. (Consider "minimum" instead of "maximum," the particular function

$$f(x,y) = x^2 + y^2,$$

and Fig. A7.)

10.43.1. A polyhedron with volume V and surface area S is circumscribed about a sphere with radius r. Then, the polyhedron can be divided into pyramids all of which have the same apex, at the center of the sphere, and the same altitude r; their bases jointly constitute the surface of the polyhedron. Therefore,

$$V = Sr/3.$$

Hence we obtain for the isoperimetric quotient

$$\frac{36\pi V^2}{S^3} = \frac{4\pi r^3/3}{V}$$

a value less than 1, since the volume of the sphere (the numerator) is obviously less than the volume V of the polyhedron circumscribed about it.

---

[7] See *Publicationes mathematicae, Debrecen,* vol. 1, 1949–50, pp. 65–70.
[8] See *Acta Mathematica,* vol. 91, 1954, pp. 143–164.

12.3.1. Apply the same checks (1) to (7) as in the solution of ex. 12.3. In the case (1) $A = 0$, $r = \infty$; the case (2) yields no obvious conclusion; the other checks work similarly as in ex. 12.3.

12.5.1. Of the checks administered to ex. 12.5, (1) and (6) turn out quite straightforward, and (7) is more obvious in the present case. If, in the case (5), the circumscribed sphere remains fixed when the tetrahedron collapses into a plane quadrilateral, the quadrilateral is inscriptible, and the vanishing of one of the factors on the right-hand side of the asserted equation yields Ptolemy's theorem.

12.5.2. We first check the expression of V in terms of the sides then that in terms of the angles.

(1) $a = b = c = \frac{\pi}{3}$: the tetrahedron is regular, $V = \frac{\sqrt{2}}{12}$.

(2) $a = b = c = \frac{\pi}{2}$, $V = \frac{1}{6}$.

(3) $a + b + c \to 2\pi$: the tetrahedron collapses, $V \to 0$.

(4) $a \to 0$, $b \to 0$, $c \to 0$: the spherical triangle is "almost plane," $V^2 \sim s(s - a)(s - b)(s - c)/9$.

(5) $A = B$: the spherical triangle is isosceles,

$$V = -\frac{\cos(A + C/2)\cos(A - C/2)\cot C/2}{3\sin^2 A}.$$

If $C \to \pi$, the tetrahedron collapses, $V \to 0$.

(6) $A = B = C$: the spherical triangle is equilateral,

$$V = -\frac{1}{12}\frac{\cos S}{\left(\sin\frac{S}{3}\right)^3}.$$

When

$$S = \frac{\pi}{2}, \frac{3\pi}{4}, \frac{3\pi}{2},$$

$$V = 0, \frac{1}{6}, 0.$$

(7) Consider a polyhedron inscribed in the unit sphere with F faces, V vertices, and E edges. If $V = n$ and all F faces are congruent equilateral triangles, then the volume of the solid consists of $F = 2(n - 2)$ congruent

tetrahedra and equals

$$\frac{(n-2)\sin\dfrac{\pi}{n-2}}{6\left[\sin\dfrac{\pi n}{6(n-2)}\right]^{3}};$$

we have used (6), Euler's formula (*cf.* ch. 3), and the fact that the area of the spherical triangle is $2S - \pi$. Our polyhedron is a regular solid. For $n = 4$, 6, and 12 we obtain the volume of the inscribed tetrahedron, octahedron, and icosahedron, respectively.

16.6.1. [Stanford 1958] (I) false. Counterexample: If in an isosceles triangle the angle opposite the base is sufficiently close to 180° all three altitudes are arbitrarily close to 0. (II) true. Even more is true. In fact, let a, m_b and m_c denote a side and the medians starting from its endpoints, respectively. Then, from a suitable triangle

$$\tfrac{2}{3}m_b + \tfrac{2}{3}m_c > a.$$

By adding this equation to the two analogous equations we find

$$m_a + m_b + m_c > 3(a + b + c)/4.$$

(III) true and can be proved by a reasoning similar to the foregoing. Yet "more" is not true: If in an isosceles triangle the angle opposite the base is sufficiently close to 0 the sum of the three bisectors is arbitrarily close to the semiperimeter.

16.6.2. [Stanford 1964] The right triangles with sides

$$3,4,5; \qquad 1,1,\sqrt{2}; \qquad \sqrt{2},\sqrt{2},2; \qquad \sqrt{3},\sqrt{3},\sqrt{6}.$$